Lecture Notes in Information Systems and Organisation 62

Lecture Notes in Information Systems and Organization—LNISO—is a series of scientific books that explore the current scenario of information systems, in particular IS and organization. The focus on the relationship between IT, IS and organization is the common thread of this collection, which aspires to provide scholars across the world with a point of reference and comparison in the study and research of information systems and organization. LNISO is the publication forum for the community of scholars investigating behavioral and design aspects of IS and organization. The series offers an integrated publication platform for high-quality conferences, symposia and workshops in this field. Materials are published upon a strictly controlled double blind peer review evaluation made by selected reviewers.

LNISO is abstracted/indexed in Scopus

Varun Gupta · Luis Rubalcaba · Chetna Gupta ·
Thomas Hanne
Editors

Sustainability in Software Engineering and Business Information Management

Proceedings of the Conference SSEBIM 2022

Editors
Varun Gupta
Multidisciplinary Research Centre for
Innovations in SMEs (MrciS)
and Department of Digital Innovations
GISMA University of Applied Sciences
Potsdam, Germany

Chetna Gupta
Computer Science & Engineering
Jaypee Institute of Information Technology
Noida, Uttar Pradesh, India

Luis Rubalcaba
Faculty of Economics, Business and Tourism
University of Alcala
Alcalá de Henares, Madrid, Spain

Thomas Hanne
Institute for Information Systems
University of Applied Sciences and Arts
Northwestern Switzerland
Olten, Solothurn, Switzerland

ISSN 2195-4968 ISSN 2195-4976 (electronic)
Lecture Notes in Information Systems and Organisation
ISBN 978-3-031-32435-2 ISBN 978-3-031-32436-9 (eBook)
https://doi.org/10.1007/978-3-031-32436-9

This Springer imprint is published by the registered company Springer Nature Switzerland AG
The registered company address is: Gewerbestrasse 11, 6330 Cham, Switzerland

Preface

An Overview

Sustainability in Software Engineering and Business Information Management volume contains the research papers presented at International Conference on Sustainability in Software Engineering & Business Information Management: Innovation and Applications (SSEBIM 2022). This conference was jointly organized by the ***University of Applied Sciences and Arts Northwestern, Olten, Switzerland***, and ***INSERAS Research Group, Universidad de Alcalá, Alcalá de Henares (Madrid), Spain*** during September 23–24, 2022. The primary goal of the conference is to bring together researchers from academia and industry as well as practitioners to share ideas, problems, and solutions relating to sustainability in software engineering, business models, and management. The target was to focus on the latest trends, techniques, and application areas adapting sustainability.

There were 112 submissions to the conference from authors all across the world. Each manuscript underwent thorough assessment at several different levels, including pre-screening by editors, reviews by at least two reviewers, and a final check by editors and a member of the international program committee. The review procedure comprised at least two review rounds; however, some articles underwent three. Ten papers were selected by the committee for presentation. Three keynote addresses by Professors **Pekka Abrahamsson** (*University of Jyväskylä, Finland*), **Leandro F. Pereira** (*WINNING Scientific Management, Portugal*), and **Lawrence Peters** (*Software Consultants International Limited, USA*) were included in the conference schedule.

This book is composed of two different but related parts, namely

- **Part A:** Sustainability in Software Engineering: Change, Growth, and Future Impact
- **Part B:** Sustainability in Business: Change, Growth, and Future Impact

The book houses the research work related to sustainability from both a technical ***(Part A)*** and business ***(Part B)*** point of view. **Part A** of the book houses the research articles which capture the technical perspective of software development companies and thus focus on sustainability in software engineering ranging from practices, tools, techniques, and methods. The software engineering community focuses on delivering software products that meet customer needs in more sustainable ways. This will capture the research articles which reflect that how software engineering teams exhibited proactiveness in their approaches to lead to the sustainable development of software that is of the highest quality and reliability. **Part B** of the book provides the business perspective on sustainability, and hence, the focus is how to make business operations more sustainable, a topic that has attracted much attention in the European Union these days. The focus is not only on the sustainability of the business, i.e., how much businesses remain competitive, but the main focus will be on human values, ethics, environment, and responsibility of the businesses ranging across multiple domains and issues. Business operations offer support to society as their corporate social responsibility, but this needs to be balanced with the flexibility and agility expected in highly competitive environments.

The chapters contained in this book range from original research articles, literature surveys, and case studies and thus provides different research reflecting the work which had been validated in real settings rather than just in laboratories.

Contributors Demographics

Authors from prestigious universities throughout the world have contributed to this book. In addition to contributing three keynote address abstracts, these authors contributed to 13 of the book's chapters. As shown visually in Fig. 1, contributors came from all over the world, including North America (United States of America (USA)), Europe (Finland, Norway, Portugal, Spain), Asia (Oman, Vietnam), and Africa (Algeria, Nigeria, South Africa).

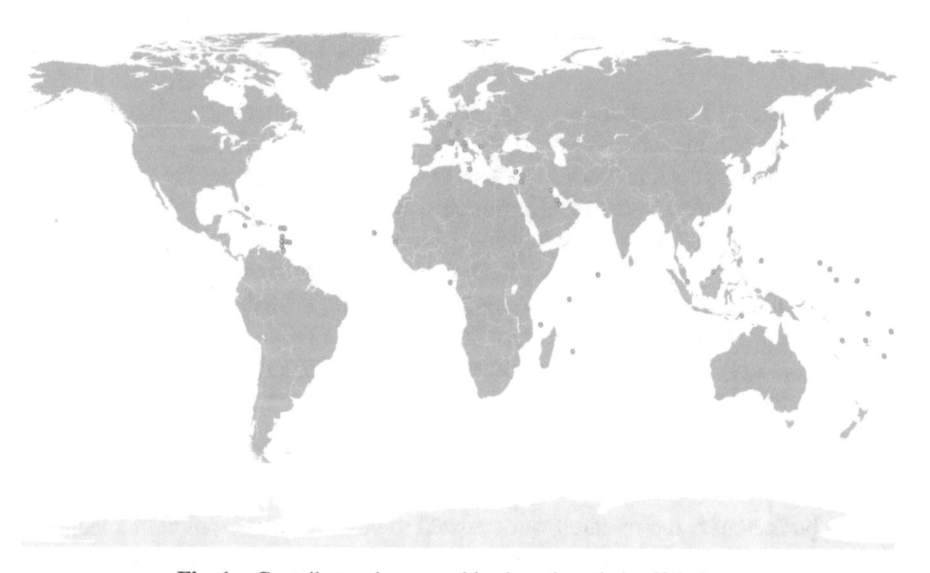

Fig. 1. Contributor demographics based on their affiliations

Organization of Book

The 13 chapters that make up this book are separated into two sections, A and B. Part A, which consists of four chapters, is focused on research on sustainability in software engineering. With its eight chapters, Part B covers the topic of sustainability in business. Despite being separate, the chapters are related to each other. The following paragraphs provide a succinct overview of each chapter.

Part A: Sustainability in Software Engineering: Change, Growth, and Future Impact.

Chapter 1 addresses the important issue of artificial intelligence (AI) fairness while analyzing software system applications. In the modern world, artificial intelligence (AI) has been applied to the same development techniques as software. There have been opinions that the evaluation of the quality of AI software should be based on the element of AI software fairness. Unfair AI software is considered shoddy software. A software system using ML/AI without a proper design, implementation, and testing might lead to severe and harmful consequences, due to its unfairness. There is already evidence of unfairness in AI systems when dealing with sensitive factors, such as race, gender, and skin color. This led to legal consequences for the organizations that develop and operate the system and practical impact on society and sustainable development. There is a lot of recent research intending to make AI software fair, accountable, and transparent. Therefore, it is extremely important to consider the issue of fairness while analyzing this kind of software. A big question is also raised. What is fair AI software? How to measure the fairness of a given AI software and how to test that fairness? The authors have mentioned three issues to answer these three questions. To achieve research objectives, the authors have compiled the definitions of fairness (especially the quantitative definitions of fairness) used by other researchers when analyzing the issue of fairness in AI software. Secondly, the authors have synthesized the approaches to solving the issue of fairness that have been proposed by many researchers, namely three stages applied to approach fairness of AI software: pre-processing, in-processing, and post-processing. In addition, the authors also outlined four methods and techniques to approach software fairness: mitigating bias; fair machine learning; learning about fair representation; fair natural language processing. Finally, it was specified what fairness definitions are used in those approaches and techniques. Thirdly, the authors have summarized the difficulties and challenges and suggested solutions to overcome those difficulties in the research on AI software fairness.

Chapter 2 focuses on developing a methodology for normative multi-agent systems. Software reliability engineering (SRE) focuses on engineering techniques for developing and maintaining software systems whose reliability can be quantitatively evaluated. Software operational profile (SOP) development is the first important step in the software reliability testing process. Several approaches have proposed methodologies to generate the software operational profile. However, almost all the proposed studies have omitted the specificities of the new software paradigms, despite that these new paradigms are generally characterized by their own concepts, methods, and methodologies. Multi-agent systems are one of these software paradigms. It will be more beneficial to propose a specific software operational profile development methodology instead of using a general one. In this chapter, the authors propose a development methodology for a specific kind of multi-agent system (called, normative multi-agent systems). This methodology is described gradually using a case study.

Chapter 3 proposes self-repair capability measurement by proposing a common Field Programmable Gate Arrays (FPGA) Partial Re-configuration (PR) self-repair process. Intelligent Embedded Systems (IES) represent a new discipline in which artificial intelligence (AI) is coupled with Embedded Systems (ESs) in order to create a new self-X-based Embedded Systems generation. Self-X capabilities, like self-repair, self-awareness, self-adaptation, etc., brought, for an ES, the ability to reason about their external environments and, as a result, adapt their behavior appropriately. Also, self-repair capability defines the fact of identifying and repairing failures in order to avoid the high cost of the ordinary repair process. FPGA provides support for implementing the self-repair capability based on the FPGA Partial Re-configuration feature. Despite that there are many approaches that proposed self-repair processes, the measurement aspects are often omitted. Indeed, the measurement allows for controlling the repair process and comparing the different approaches. In this chapter, the authors discuss the self-repair capability measurement by proposing a common FPGA PR self-repair process followed by a set of hardware (HW) and software (SW) metrics dedicated particularly to FPGA PR-based IES.

Chapter 4 focuses on investigating female students' participation in software engineering projects to support gender-aware course optimization. Among the 17 Sustainable Development Goals (SDGs) established by the United Nations in 2015, gender equality is an important goal that closely relates to the academic context. Engineering is a male-dominated field, both in terms of the gender majority and the way in which engineering tasks are framed and valued. The objective of this work is to enhance the understanding of female students' participation in software engineering projects to support gender-aware course optimization. Since 2015, the authors have investigated the activity profiles of female students in terms of software engineering activities in a fourth-year software project course. Empirical evidence has been collected through surveys, structured interviews, and project reports from 39 projects. We found that the active activity areas of female students are project management and requirement engineering, while the areas lacking active involvement are architecture and Scrum methodology. While the findings differ from those of some previous studies, they suggest which course and project settings will facilitate the active participation of females in such project courses.

Part B: Sustainability in Business: Change, Growth, and Future Impact.

Chapter 5 presents an exploratory case study to investigate the sustainability challenges for a large software technology company that operates platforms and digital services. Sustainability is defined as a development that takes into consideration both current and future needs, based on three so-called pillars: economic, environmental, and social needs. In practice, it is not trivial for companies to simultaneously include all these aspects in their development strategies. Large companies are increasingly faced with the dilemma of balancing their economic performance and sustainable environmental and social impact. The recent development of information and communication technology (ICT) has certain implications for achieving cooperates sustainability goals. Although both the theory and literature on sustainable development are rich, there is a gap in empirical cases that provide practical insights into the challenges for companies, particularly the technology-focused ones. Our objective is to explore the sustainability challenges for a large software technology company that operates platforms and digital services. We conducted an exploratory case study based on various sources of information. By matching the company's sustainability goals with the observed initiatives and product characteristics, we revealed insights into five different challenges at the organizational level. The study has implications for both research and practice on sustainability dilemmas in the ICT sector.

Chapter 6 investigated the Information Technology (IT) opportunities in South Africa. In this chapter, the authors tried to answer three questions, namely which IT jobs are in the highest demand in South Africa; which companies advertised the most job vacancies in IT in South Africa; which locations (provinces and cities) had the largest number of job vacancies in IT in South Africa?. The authors analyzed 3,355 jobs in Information Technology (IT) published on LinkedIn's South Africa website with the aim of identifying the leading locations for such jobs, as well as the recruitment companies. The results indicate that three (of the nine) provinces accounted for 91 percent of all IT job vacancies in South Africa, namely Gauteng (46.7%), Western Cape (39.4%), and KwaZulu-Natal (4.8%). Analysis of job distribution by cities revealed that Cape Town was home to 33 percent of all IT jobs in South Africa, followed by Johannesburg (24%) and Pretoria (3%). The Northern Cape province had the fewest IT job openings, followed by the Free State and Limpopo. Overall, the three most in-demand IT jobs were DevOps Engineers, Software Engineers, and Data Engineers. Among the companies with the largest number of IT jobs, DigiOutsource came first, followed by takealot.com and impact.com. Recruitment firms advertised the biggest percentage (50%) of the top IT jobs in South Africa. The research findings will add to the existing literature on IT opportunities in South Africa. Given that location is an essential part of the decision-making process among individuals considering job offers, our findings will provide applicants with a good understanding of how job opportunities in IT vary across locations in South Africa.

Chapter 7 analyzed the main impacts of COVID-19 on the hospitality business and proposes a generally reinvented business model for hospital industry innovations. Currently, the hospitality business is facing severe challenges due to the COVID-19 pandemic, forcing hotels to quickly adapt to a new operational reality and address in a critical way the sustainability challenge. Such adaptation requires redefining the current business models since they are not adequate for the present global scenario. The study's primary goal is to analyze the main impacts of COVID-19 on the hospitality business and propose a generally reinvented business model, aiming to contribute to the hospitality industry's reinvention in the global scenario. A literature review was developed to fulfill these objectives, and an in-depth interview was conducted with 14 professionals who work at different hotels. The obtained results demonstrated the reinvention of the hotels' business model must include significant investment in technology and digital communication since these two are vital for the industry's growth and evolution by providing customers with a sense of trust.

Chapter 8 pertains to software project management and provides useful directions for improving the current practices. Software project management is a topic that is rarely found in the call for papers at software engineering conferences. This combined with the fact that most software project managers have not been trained in management methods has resulted in software project managers adopting methods that they believe are beneficial but are not. This is not a sustainable situation as software projects are becoming more expensive with far-reaching consequences for failure. This paper cites some of the more common fallacious methods and provides published references refuting their presumed benefits together with suggestions as to how to improve the current state of affairs.

Chapter 9 explored through qualitative study the communication between information systems in Portuguese hospitals. In the health system, the adoption of Informatics and Communications Technologies (ICTs) is becoming visible, and with that more heterogeneity and interoperability between Information Systems (IS) and medical data. The technological revolution enhances interoperability between Health Information Systems (HIS) as a new challenge to achieve sustainability. Thus, the goal of this study is to explore and analyze the communication between Information Systems in Portuguese hospitals. This study follows qualitative research, sixteen semi-structured interviews were conducted with healthcare professionals and hospital suppliers to collect their experience on interacting with HIS, the stakeholders' role, and the measures that are considered crucial to improve the current paradigm. The subsequent analysis was supported with the MAXQDA program to organize and identify the keywords, explanations, and views presented by the interviewees. The key findings of the study were the inadaptation of public hospitals' boards to manage HIS and promote sustainability, the importance that healthcare professionals have in the implementation of new HIS in hospitals, and the lack of interoperability between the systems resulting in slower and more complex navigation for the user. To achieve successful interoperability between HIS, hospitals need to have a clear hospital strategy focused on the digitalization of processes. Providing an IT structure that supports the implementation and management of the systems, managing the relationship with the hospital's suppliers, and encouraging the healthcare professionals to involve and use this technology. Further research needs to consider a larger sample and other research methods to identify new constraints and opportunities about this topic.

Chapter 10 provides interesting information about the role of Information and Communication Technology (ICT) in business management. A Great Resignation is going on and workers are leaving their jobs at rates never seen before. They do it under economic circumstances never experienced by a whole generation of workers: the highest inflation rates in more than 40 years, in the presence of the longest period of low-interest rates, and with apparently overvalued stocks and real estate markets. Not surprisingly, in such a scenario, political discontent arises, leading to an unprecedented success of populists' proposals at political elections. To this date, mainstream theories at economic, business management, and political levels appear unable to fix the multiple existing economic and social problems. In this paper, it is argued that the use of information and communication technologies provides new perspectives for business management, in terms of relocation of workers at different geographic settings, but also establishing different contractual situations with their (former) employers and performing different functional-related tasks at the value chain level. That is, uncertain times bring new opportunities for younger workers, who may opt to live as digital nomads, locating their residences where fewer taxes need to be paid. Renouncing to work for a company, they can establish their selves as freelancers, self-employed, or entrepreneurs, with the aim of offering their services to companies or individuals located anywhere, while developing a more nuanced knowledge of companies and industries. Developing these new business management perspectives requires, however, a deeper understanding of how entrepreneurial intentions relate to other factors, such as political positioning, human values, or religious feelings.

Chapter 11 provided an interesting evaluation of Information Technology (IT) in the construction industry in the Sultanate of Oman. The use of Information Technology (IT) in construction projects is essential for the development of the company's works. This study aims to evaluate the use of IT in the construction industry in the Sultanate of Oman. It also seeks to identify the programs primarily used in the construction industry, the factors that affect the use of IT, and the approaches to overcoming the barriers to using this technology in construction projects. Initially, an interview was conducted with professional engineers and contractors in the selected construction industry and the Ministry of Housing related to implementing IT. Then, an electronic questionnaire was completed, and the data collected from the questionnaire were analyzed with the Statistical Package for Social Sciences (SPSS). In general, the analysis showed that all interviewees had a clear understanding of the importance and scope of IT in the construction industry; however, there were still several barriers to its implementation. The results showed that the most common obstacle that reduces the use of IT in the construction industry in the Sultanate of Oman is the weak capabilities of employees in the use of IT. Also, it was found that the primary strategy that might help reduce the barriers to using IT in construction projects is to intensify courses in the use of IT. Finally, the results of the study might help the decision-makers in ministries and construction projects in the Sultanate of Oman toward improving the construction sector.

Chapter 12 highlights the current return on investment (ROI) practices in training projects practices used by organizations and also provided recommendations and directions for future research. Although the advanced practices on how to estimate and measure business initiatives benefit, it is still an uncommon practice to apply this knowledge and methodologies on HR initiatives, in particular, training projects. Knowing how to measure the ROI in training projects is becoming a critical skill for HR executives as they need to justify whether that investment was effective and whether there was any return generated with the training projects to their organization's employees and ultimately to the business goals. This research presents the main results from research about the current ROI in training projects practices used by organizations, where it was possible to conclude that levels 1 and 2 (reaction and learning evaluation) are frequently used; however, the remaining levels (impact, application, and ROI analysis) are often neglected. Finally, the article also points out recommendations and directions for future research on the topic.

Sustainability in software engineering refers to the creation and upkeep of software systems in a way that is ethical and morally righteous, commercially viable, and technologically practicable. This covers things like making sure software systems are long-lasting and still usable, eliminating waste, fostering diversity and inclusion, and lowering carbon emissions. The effects of software on society and the environment have drawn considerable attention in recent years, which has caused the software engineering sector to put more emphasis on sustainability. Given that sustainability in software engineering affects not only the technology sector but also the global economy and society as a whole, its influence is anticipated to continue to grow in the future.

When it comes to artificial intelligence (AI), sustainability in software engineering also entails making sure that AI systems are created and used in a way that encourages fairness, openness, and ethical decision-making. This encompasses factors like minimizing detrimental effects on marginalized communities, mitigating bias and discrimination in AI systems, and fostering diversity in AI development and deployment. In terms of female participation, sustainability in software engineering also entails fostering inclusion and diversity in the tech sector, which includes raising the proportion of women working in leadership positions in the fields of software engineering and AI. In addition to ensuring that women's opinions and experiences are included in the creation and application of technology, this can assist in addressing gender gaps in the sector.

The term "normative multi-agent systems" refers to systems where a set of guidelines is defined to direct autonomous agents' behavior. Normative multi-agent systems can be used in the sustainability context to make sure that the conduct of autonomous agents in a particular system complies with sustainability concepts and objectives. For instance, a normative multi-agent system may be employed to control energy usage in a smart grid, with autonomous agents serving as various grid-connected devices and appliances. The system's set rules and standards could guarantee that the gadgets and appliances run as efficiently and with as little loss of energy as possible. Additionally, normative multi-agent systems can be used to make sure that autonomous agents' conduct encourages justice and moral decision-making in AI systems, as well as diversity and inclusion in the development and application of technology. Normative multi-agent systems can help to ensure sustainability in AI and software engineering by offering a framework for developing rules and norms that are consistent with sustainability concepts and objectives.

The term "self-repair capability measurement" refers to a system's capacity to locate and fix errors or problems on its own, without the assistance of a third party. Self-repair capability can be crucial in maintaining the longevity and ongoing utility of software systems in the context of sustainable software engineering. A software system with strong self-repair capabilities, for instance, can identify and fix problems or failures more quickly and effectively, eliminating downtime and the need for pricey maintenance and repairs. In addition to assuring the software system's continuing operation and usefulness, this can assist limit waste and the software system's carbon footprint. Additionally, the ability to self-repair can help to increase the security and dependability of software systems, lowering the danger of data breaches and other security problems. In order to advance sustainability in software engineering and AI, this can assist assure the protection of sensitive data and the privacy of users. Self-repair capacity measurement is a crucial component of sustainability in software engineering because it ensures the robustness, security, and lifespan of software systems while lowering the environmental impact and waste generated by their upkeep and repair.

In summary, *sustainability in software engineering* is a broad notion that includes social, economic, environmental, and technical factors. It is critical to ensure that sustainability concepts are incorporated into AI development and deployment as it continues to play an increasingly significant role in our society to promote justice, transparency, and ethical decision-making. The development and deployment of software systems, as well as the environment and society at large, are all impacted by sustainability in the software engineering field. We can make sure that technology is created and used in a way that supports a more sustainable future by incorporating sustainability concepts into software engineering methods and processes.

Sustainability in business refers to the incorporation of environmental, social, and economic factors into corporate strategy and day-to-day operations. This involves initiatives to reduce adverse environmental effects, advance social and economic well-being, and guarantee long-term financial viability. Growing awareness and concern about how company operations affect the environment and society in recent years has caused the business sector to place more emphasis on sustainability. Companies are adopting sustainable practices and activities, such as lowering carbon emissions, promoting ecologically friendly goods and services, and encouraging diversity and inclusion in the workplace, as they increasingly understand the value of sustainability for their long-term success. As more businesses embrace sustainable practices and customers become more aware of the environmental and social impact of the goods and services they buy, sustainability in business is set to have a large impact in the future. The corporate world will continue to adapt and flourish as a result of governments and regulatory organizations taking sustainability into account in their policies and laws.

It is crucial to research the sustainability issues facing big software companies. This is due to the fact that they affect the environment, social responsibility, financial viability of the organization, and best practices and novel solutions that may be used more widely to advance sustainability in the corporate world.

Jobs in Information Technology (IT) face sustainability issues relating to the environmental impact of computing and the social impacts of the digital divide, particularly in South Africa. IT businesses in South Africa must take into account both the environmental impact of the goods and services they offer as well as the energy use and carbon emissions connected with their operations. This involves considering the data centers' and computing equipment's energy efficiency as well as the supply chain's sustainability for creating and distributing IT goods and services. IT businesses in South Africa must take into account the societal implications of the digital divide and unequal access to technology and the Internet. Guaranteeing that everyone has access to the advantages of technology involves tackling concerns related to affordability, connectivity, and digital literacy. In conclusion, there are significant sustainability issues affecting South African IT occupations that have an impact on both the environment and society. These issues must be considered by businesses and individuals operating in the IT sector when making decisions and working to advance sustainable technology and universal access to the Internet.

The COVID-19 pandemic has had a substantial negative effect on the hospitality sector's sustainability efforts, with many firms now confronting formidable waste management, reduced sustainability efforts, changes in consumer behavior, and a rise in the usage of online and delivery services. Businesses in the hospitality industry will need to reassess their sustainability plans and embrace fresh tactics that support sustainability in a post-pandemic world in order to mitigate these problems.

There are many myths about software project management, a crucial component of software development. It is necessary to recognize the myths about software project management and finally dispel them. Additionally, recognizing the significance of sustainability in software development can aid in the promotion of more sustainable software development methods. This can be accomplished by focusing on the long-term effects of software development operations on the environment and society and by having a more sophisticated knowledge of software project management.

In Portuguese hospitals, efficient information system communication is essential for maintaining the long-term viability of healthcare operations. Improved information system communication can significantly contribute to the promotion of sustainability in Portugal's healthcare industry by fostering accurate and effective data management, higher effectiveness, better patient outcomes, and less environmental impact. Understanding the effects that healthcare operations have on the environment and society, as well as supporting more sustainable healthcare operations that are prepared to face future difficulties, requires a study of information system communication in Portuguese hospitals.

For a corporation to comprehend the possible threats to its workforce stability, business continuity, financial stability, and reputation, it is crucial to analyze mass resignations and their effects on sustainability. Businesses can create plans to stop mass resignations and keep a viable workforce by being aware of these hazards.

The construction sector depends heavily on Information Technology (IT), which also has the potential to have a big impact on sustainability. Understanding how technology can be utilized to improve sustainability in construction operations requires studying IT in construction and sustainability. It is feasible to create strategies for leveraging technology to use it to boost productivity, improve decision-making, improve communication, lower costs, and assist sustainable design by understanding the function of IT in construction.

Businesses should take the return on Investment (ROI) of training initiatives into account, especially in terms of sustainability. Understanding the advantages of investing in staff development and encouraging sustainability in corporate operations requires research on the return on investment of training projects and sustainability. It is feasible to build strategies for investing in employee training and promoting sustainability in the long run by knowing the ROI of training projects.

Sustainability in business is a crucial component of the world economy and society and will have a significant impact on how business is shaped in the future. Businesses that put a strong emphasis on sustainability will be better positioned to compete in a market that is changing quickly and becoming more mindful.

This book contains intriguing articles on sustainability in business management and software engineering, with cross-disciplinary applications. Promoting sustainable practices in a technology-driven society requires researching sustainability in software engineering and business management. It is possible to create strategies for increasing sustainability in software engineering and business management by studying the effects of technology on the environment and on business operations.

The book's editors wish readers to experience a satisfying and enlightening interaction with it. We anticipate that readers will find the book's contents to be interesting, understandable, entertaining, transforming, and worthwhile. Happy reading and learning to all of our readers!

Varun Gupta
Chetna Gupta
Thomas Hanne

Organization

Honorary Chairs

Rolf Dornberger	School of Business, Institute for Information Systems, FHNW University of Applied Sciences and Arts Northwestern, Basel, Switzerland
Luis Rubalcaba	Director (INSERAS research group) & Professor (Economic Policy, Department of Economics and Business Administration), University of Alcala, Spain

Organizing Committee

Thomas Hanne	School of Business, Institute for Information Systems, FHNW University of Applied Sciences and Arts Northwestern, Olten, Switzerland
Varun Gupta	Head (Multidisciplinary Research Centre for Innovations in SMEs (MrciS)) & Professor of Digital Innovations, GISMA University of Applied Sciences, 14469 Potsdam, Germany
Chetna Gupta	Jaypee Institute of Information Technology, Noida, India

International Program Committee

Adam Sulich	Wroclaw University of Economics and Business, Poland
Amir Karbassi Yazdi	Islamic Azad University, Iran
Andy Borchers	Lipscomb University, USA
Anh Nguyen-Duc	University of South-Eastern Norway, Norway
Asta Valackiene	Mykolas Romeris University, Lithuania
Basim Al-Najjar	Linnaeus University, Sweden
Bedilia Estrada-Torres	Universidad de Sevilla, Spain

Daniela S. Cruzes	Norwegian University of Science and Technology (NTNU), Norway and SINTEF Digital, Trondheim, Norway
Dariusz Milewski	University of Szczecin, Poland
Deniz Cetinkaya	Bournemouth University, UK
Giuseppe Desolda	University of Bari, Italy
Hironori Washizaki	Waseda University, Japan
Hongji Yang	Leicester University, England
Jody Fry	Texas A&M University-Central Texas, USA
José Arzola Ruiz	Havana Technological University "José Antonio Echeverría" (CUJAE), Havana, Cuba
Lawrence Peters	Software Consultants International Limited, USA
Leonardo Salvatore Alaimo	Italian National Institute of Statistics-Istat, Italy
Leandro Ferreira Pereira	Winning Scientific Management, Portugal
Mathew (Mat) Hughes	Loughborough University, UK
María Belén Usero Sánchez	Universidad Carlos III de Madrid, Spain
Mariantonietta Fiore	University of Foggia, Italy
Martin Gilje Jaatun	University of Stavanger, Norway and SINTEF Digital, Trondheim, Norway
Nadia Di Paola	University of Naples Federico II, Italy
Nuno M. Garcia	University of Beira Interior, Portugal
Peiyi Zhu	Changshu Institute of Technology, China
Rafael Capilla	Universidad Rey Juan Carlos, Spain
Serje Schmidt	Feevale University, Brazil
Sunil Vadera	University of Salford, Greater Manchester, England
Virginia Hernández	Universidad Carlos III de Madrid, Spain
Yan Luo	University of Massachusetts Lowell, USA

Abstracts of Keynote Addresses

Mission Impossible? Implementing Ethical AI Systems (Keynote Address (Academia))

Pekka Abrahamsson

Tampere University, Finland

There is a common agreement that ethical concerns are of high importance when it comes to systems equipped with artificial intelligence (AI). Demands for ethical AI are declared from all directions. As a response, in recent years, public bodies, governments, and universities have rushed in to provide a set of principles to be considered when AI-based systems are designed and used. We have learned, however, that high-level principles do not turn easily into actionable advice for practitioners. Hence, also companies are publishing their own ethical guidelines to guide their AI development. These guidelines do not seem to help the developers. To bridge this gap, we present a method for implementing AI Ethics in practice. The ECCOLA method has been developed in collaboration with researchers and practitioners in the field, and it is under proof-testing in several AI companies. The presentation outlines the method and its practical use cases.

Software Project Management: Myths Versus Reality (Keynote Address (Industry))

Lawrence Peters

Software Consultants International Limited, USA

While there have been some projects deemed successful, software engineering has a long history of projects being late, over budget, and canceled. This is in spite of a constant stream of new programming methods, programming languages, and software project management "systems" all promising to solve the "software problem." This talk attempts to put this history into perspective by exploring the importance of the software project manager and its role in bringing software projects to a successful conclusion. As part of this presentation, several beliefs software project managers hold as self-evident are examined and put to the test of reality from published studies showing contrary results. Recommendations for improving this situation via training are included showing the need as a means of reducing the creation of beliefs based on hearsay but based on facts and data.

Strategic Problem-Solving for Reinventing Your Business Model (Keynote Address (Industry))

Leandro F. Pereira

WINNING Scientific Management, Portugal

The business world is rapidly progressing into a new era, in which the prevalence of a common purpose is the key criterion for ecosystems when passing judgment on an organization. Stakeholders have become increasingly demanding and fully alert to such matters as the quality of the interaction with an organization, valuing their experience and expectations, while assessing organizational conduct and behavior. The pandemic has greatly accelerated the advent of Industry 4.0. Business models have sought to strengthen their technological dimension, supply chains have been restructured, and education and academia have needed to reinvent themselves fully. Many companies and businesses were forced to diverge from their incumbent strategies to more innovative ones, adapting to a new reality in order to survive. It has become abundantly clear that the only truly sustainable competitive advantage for an organization lies in its capacity to reinvent itself and replace incumbent paradigms with better business models, ones that best fit a dynamic and constantly changing macro-environment. It is also key to emphasize the role of technology and its omnipresence across all types of businesses, rendering the notion of business reinvention meaningless without the inclusion of a technological paradigm shift. Digital transformation has majorly impacted the way brands operate and manifest themselves next to key constituencies and, most remarkably, consumers. It is rather cumbersome to conceive of the idea of selling products or providing services, short of technology being present at some stage of the journey. In particular, the use of platforms or sales systems has become almost a requirement for those who want to explore the online market. In recent years, e-commerce and e-marketplaces have become the primary means for purchasing products and services, but this was nothing compared to the magnitude of what we are witnessing today. People want to be surprised, they want new and different things, and they want them now. They seek ease of access and simplification of their daily tasks and routines. Perhaps it is not the product or the service that should be changed, but the way in which the user/consumer experience utilizes the product or service. It is important to note that company lifecycles have never been shorter, as a consequence of a context of V.U.C.A (volatility, uncertainty, complexity, and ambiguity), so a permanent transformation of the business model is a mandatory principle for survival. Digital business strategy and innovation are thus mandatory for the creation of a sustainable competitive advantage.

Contents

Brief CVs of Editors

Prof. Dr. Varun Gupta is Professor at GISMA University of Applied Sciences, Potsdam, Germany. He also serves as Head of the Multidisciplinary Research Centre for Innovations in SMEs (MrciS) at GISMA. He has maintained various associations with Universidad de Alcalá in a number of capacities, most notably as Postdoctoral Researcher, Researcher, and Co-Director of a master's degree in Innovation Economics, Management, and Technology. He was also Visiting Postdoctoral Researcher with the Software Engineering Research Group (SERG), Department of Computer Science, Lund University, Sweden. Also, he was a visiting scholar at the University of Toronto, Toronto, Canada.

Professor Gupta was also Researcher with Sapienza University of Rome, Italy, Free University of Bozen-Bolzano, Italy, University of South-Eastern Norway, Norway, and Poznań University of Technology, Poland. He has also been Visiting Professor at Uniwersytet Szczeciński, Poland.

Additionally, he has a substantial level of professional experience working in several firms in the fields of software engineering, innovation management, and globalization. He holds a Doctorate (cum laude) in Organizational Engineering as well as in Computer Science and Engineering. He earned a Bachelor of Technology (Hons.), and a Master of Technology (By Research) in Computer Science & Engineering. He also holds an MBA and a Máster en Dirección Internacional de Empresas.

He is Associate Editor of IEEE Access (published by IEEE, and SCIE indexed with a 3.367 impact factor), PeerJ Computer Science (published by PeerJ, SCIE indexed with a 2.41 impact factor), PLOS One (SCIE indexed with a 3.752 impact factor), International Journal of Computer Aided Engineering & Technology (published by Inderscience Publishers, Scopus indexed), IEEE Software blog, and Journal of Cases on Information Technology (JCIT) and is Former Editorial Team Member of the British Journal of Educational Technology (BJET) (published by Wiley, SCIE indexed).

He has worked on multiple projects that have been supported by various agencies such as the European Union and the Spanish National Program. He is Principal Investigator (PI) of the "Technology-based Global Business Model Innovation" project funded by Winning Scientific Management, Portugal amounting to 18K Euros. He has authored numerous research articles in top-rated, high-impact factor, and highly ranked (Q1 and Q2) international journals (Journal Citation Reports (JCR) and/or Scopus). He is Author/Editor of 16 books which have been published by top publishers like Springer, Taylor & Francis, and IGI Global. He had supervised one doctoral and 50 master's students.

He was awarded the "Best Editor Award" by Inderscience Publishers for his contributions to the International Journal of Computer Aided Engineering & Technology (Scopus Indexed) as Associate Editor. His area of interest is evidence-based software engineering, innovation management, digital transformations and innovation, technology adoptions in SMEs, entrepreneurship, and international business management.

Dr. Chetna Gupta is Professor in the Department of Computer Science & Engineering and Information Technology at the Jaypee Institute of Information Technology, Noida, India. She has more than 17 years of experience in research and teaching. She is currently serving as Associate Editor, in the International Journal of Computer Aided Engineering and Technology (Inderscience) and the Journal of Cases on Information Technology (IGI Global-USA). She has served as Guest Editor of many special issues published/ongoing with leading international journals and is also editing books to be published by IGI Global and Taylor & Francis (CRC Press). Her area of interest includes software engineering, cloud computing, blockchain technology, and applications of machine learning and data mining.

Prof. Dr. Thomas Hanne received master's degrees in Economics and Computer Science and a PhD in Economics. From 1999 to 2007, he worked at the Fraunhofer Institute for Industrial Mathematics (ITWM) as Senior Scientist. Since then, he is Professor for Information Systems at the University of Applied Sciences and Arts Northwestern Switzerland and Head of the Competence Center Systems Engineering since 2012.

Thomas Hanne is Author of more than 200 journal articles, conference papers, and other publications and Editor of several journals and special issues. His current research interests include computational intelligence, evolutionary algorithms, metaheuristics, optimization, simulation, multicriteria decision analysis, natural language processing, machine learning, systems engineering, software development, logistics, and supply chain management.

Sustainability in Software Engineering: Change, Growth, and Future Impact

Fairness Requirement in AI Engineering – A Review on Current Research and Future Directions

Nga Pham[1,2], Hung Pham-Ngoc[2], and Anh Nguyen-Duc[3](✉)

[1] Dainam University, Hanoi, Vietnam
ngaptt@dainam.edu.vn
[2] VNU University of Engineering and Technology, Hanoi, Vietnam
hungpn@vnu.edu.vn
[3] University of South Eastern Norway, Bø i Telemark, Norway
angu@usn.no

Abstract. Currently, Artificial Intelligence (AI) has been applied to the same development techniques as software. There have been opinions that the evaluation of the quality of AI software should be based on the element of AI software fairness. An unfair AI software is considered shoddy software. There is a lot of recent researches intending to make AI software fair, accountable and transparent. Therefore, it is extremely important to consider the issue of fairness while analyzing this kind of software. A big question is also raised. What is fair AI software? How to measure the fairness of a given AI software and how to test that fairness? This paper will summarize the concepts of fairness in AI software that have been introduced as well as the method of measuring and testing fairness in AI software according to those concepts. Based on an ad-hoc literature review, we summarize some recent findings in the area of requirement engineering for AI fairness and point out some research gaps.

Keywords: Fairness · Bias · AI · Definitions of Fairness · AI Software Fairness · Literature review

1 Introduction

The applications of Machine Learning (ML)/Artificial Intelligence (AI) are gradually affecting the industries, society and our individual lives in both positive and negative manners. A software system using ML/AI without a proper design, implementation and testing might lead to severe and harmful consequences, due to its unfairness [1]. There is already evidence of unfairness in AI systems when dealing with sensitive factors, such as race, gender, skin color, etc. [2]. This led to not only legal consequences for the organizations that develop and operate the system but also practical impact on society and sustainable development [3, 4]. Recent researches have revealed that the majority of AI/ML-based software systems have not sufficiently dealt with fairness [5].

V. Gupta et al. (Eds.): SSEBIM 2022, LNISO 62, pp. 3–13, 2023.
https://doi.org/10.1007/978-3-031-32436-9_1

Modern Software Engineering (SE) researchers are looking at AI fairness from an engineering aspect, and considering fairness as a quality attribute of an AI software system [1, 6–8]. It is claimed that AI/ML-based systems can only be trustworthy if their fairness can be verified and validated. In the first step, we need a systematic approach to define AI fairness in the scope of the developing system, describe fairness as an AI software attribute to be integrated into later phases of AI software development processes, track the development of AI fairness and be able to verify it during the soft-ware development and operation. However, without a theoretical and systematic approach, researchers and designers might lack a road map for crafting interventions to help promote fairer AI systems [9].

In this paper, we conducted a semi-structured literature review about fairness and AI from a software engineering perspective. We investigated existing definitions and classification of AI fairness, current approaches to deal with AI fairness, and challenges in engineering fairness in large software systems. Although our limitation on time and effort does not allow us to do a systematic literature review, our research has covered the highly cited papers as well as the recent secondary studies on AI ethics and fairness.

The paper is organized as follows. Section 2 provides the background of requirement engineering, fairness as requirement attribute, and ensuring fairness in AI-based software systems. Section 3 presents the research directions with current knowledge and future direction. Section 4 concludes the paper.

2 Background

2.1 Probabilistic Requirement Engineering

In a traditional SE process, requirement engineering includes requirements elicitation, requirement analysis, requirements specification, and requirements validation. Figure 1 depicts the technical process of software requirements activities. In that process, *requirement elicitation* concerns gathering and extracting requirements from relevant stakeholders. *Requirement analysis* deals with revealing detailed characteristics of the gathered requirements, their relationships, and their connections to stakeholders and environments. *Requirement specification*, in particular, formal software specification is about using formal language, template, mathematical formulation to define the requirements and their constraints. Values such as fairness are abstract notions and have to be made more concrete into requirement specification. Eventually, requirements can be validated against stakeholders' needs or verified if the implemented software be-haves according to its specification.

In many application domains, most of the requirements have a probabilistic verification technique [10]. For safety, the probability that a certain hazard occurs must be less than the probability allowed for such a system. For fairness, the probability that an AI system mistreats a sensitive group should be reduced to a certain level. Because of such characteristics, there is a tendency to use probabilistic verification techniques in verifying non-functional requirements of a software system in general and of an AI software in particular (including requirements on the fairness of the software). In addition, probabilistic verification techniques that work on models in addition to deterministic and non-deterministic decisions allow probabilistic decisions [10]. But in order to apply

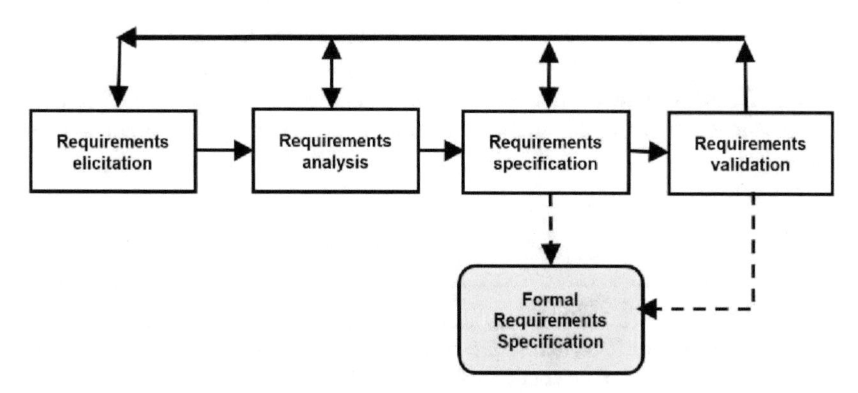

Fig. 1. Requirement Engineering activities

probabilistic verification techniques, the necessary properties must be precisely specified in terms of probability logic [10].

2.2 Fairness as a Requirement Attribute

There is no universal definition of fairness, which shows the difficulty in solving this problem [5]. In a general way, fairness can be understood as follows, "everyone enjoys equal rights and benefits in the same situation, a certain aspect" [5]. Formal definitions of fairness have emerged, and researches have exploded on it in various forms since the 1960s. In the 1960s, issues of fairness were often predicated on the identification of unfair criteria (bias, discrimination, injustice, inequality, etc.) [11]. Robert Guion has defined unfair discrimination as "when people with equal probability of being successful in a job have an unequal probability of being employed for that job" [11]. However, he recognized the challenges in using this definition. By the 1970s, there was a shift in the view of researchers on the issue of fairness that was moving away from defining unfair criteria to possible fair criteria [11]. Table 1 describes several categories of fairness that we found from the literature.

In addition to the philosophical and ethical debates about definitions of fairness, formulating a generic concept of fairness is challenging. Indicators often emphasize individual fairness (e.g., everyone is treated equally) or group fairness, where it is more clearly distinguished from within groups (e.g., women versus men) and subgroup fairness (e.g., young women and black men). Currently, the combination of these ideals using established definitions has been proven to be mathematically correct [8]. However, it is a fact that the formal definitions of fairness have not yet reached a consensus among scholars.

When looking at algorithmic fairness, it is quite straightforward to understand: looking for ensuring that the outcome of a classifier is not biased towards certain values of sensitive variables such as age, race, or gender. But what about the fairness of software? An example of a student loan processing software is described in Brun et al. [26]. Based on group fairness, the same portion of applications of each race should get loans. However, at the individual or procedural level, it states that no two applicants identical in every way except race should result in one receiving the loan and the other not. So if there are two applicants who differ only in race, they should either both get a loan or neither of them should get a loan. Both of these properties seem desirable in a fair system. However, when applied to the real world, the two definitions cannot be satisfied simultaneously.

Table 1. Types of fairness

Type of Fairness	Definition	Ref.
Organiza tional fair ness	***Distributive fairness***: Fairness with respect to the allocation of outcomes such as pay and other resources ***Procedural fairness:*** Fairness defined by the process employed to reach or decide an outcome ***Informational fairness***: Fairness defined by how employ ees are treated by their organization: Degree that information is provided to help employees understand processes taken to achieve fairness and their outcomes ***Interpersonal fairness***: Fairness defined by how employees are treated by their organization: Degree that employees are treated with respect and dignity	[9]
Unit-level fairness	***Group fairness:*** Some form of statistical parity (e.g. between positive outcomes, or errors) for members of different protected groups (e.g. gender or race) ***Individual fairness***: People who are 'similar' with respect to the classification task receive similar outcomes.	[25]
Domain-based fairness	Interactional fairness: Fairness among actors/stakeholders Procedural fairness: Fairness in rules, regulations that de fines the outcome of a procedure Outcome fairness: The distribution of the outcome (grades, loan, salary, etc.)	[9]

Fairness, like other properties of a system, should be explored from talking with customers. There needs to be a guideline that supports the elicitation and specification of fairness requirements, i.e. to identify the type of fairness that is desirable, the context where the fairness should be employed, and assumptions from the customers regarding the fairness requirements [26]. Analysis and prioritization are also needed for fairness requirements. Combining different types of fairness might lead to inconsistent and unsatisfiable requirements. Analyses that help understand the possible implications of such constraints and how fairness requirements affect other requirements are critical for understanding the trade-offs and for properly specifying such requirements.

2.3 Ensuring Fairness in AI-Based Systems

Technical approaches to ensuring fairness in AI software are usually applied in three different stages: pre-modeling (*Pre-processing*); at the point of modeling (*In-pro-cessing*) or after modeling (*Post-processing*), i.e., they emphasize intervention [7]. A common framework for intervention-based approaches to fairness in machine learning software is proposed by [5]. Figure 2 depicts a model for an intervention-based approach to ensuring fairness in ML. While not all approaches to fairness in AI software fit into this framework, it provides a consistent approach to fairness in AI software.

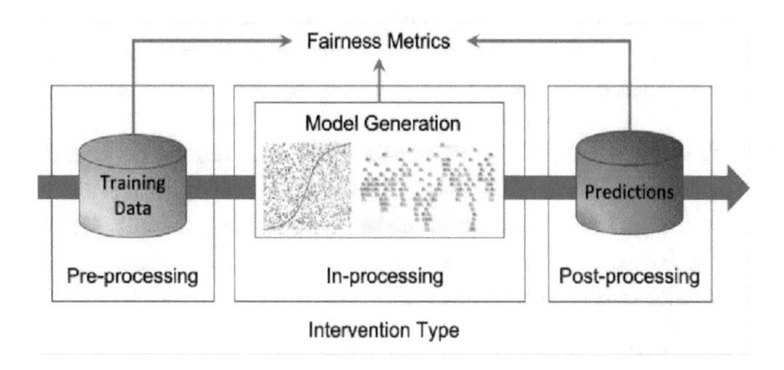

Fig. 2. Illustration of equitable intervention in machine learning [8].

Stage 1: Pre-processing
One problem mentioned in this approach is that bias, discrimination or unfairness occurs in the data itself and the distribution of sensitive variables [14]. Therefore, The *Pre-processing* stage tend to change the sample distribution of protected variables, or in general, perform specific transformations on the data with the aim of eliminating discrimination against the protected variables processed from the training data [8]. The main idea here is to train a model on the "corrected" dataset. Pre-processing is arguably the most flexible part of the data science pipeline, as it makes no assumptions on the choice of modeling technique to be applied afterward [8].

At this stage, there have been many articles that delve into research on bias and fairness in different fields and sub-fields of artificial intelligence. Various methods have been proposed to meet some of the definitions of fairness, such as:

Bias Mitigating: To minimize the impact of bias in data, several general approaches have been proposed in favor of being well prepared when using data [5, 6, 8] while preventing discrimination in data mining, targeting direct, indirect, and concurrent impacts.

Fair Community Detection: With the community detection method Ninareh Mehrabi and colleagues have shown the existence of these bias patterns, they propose a new community detection method CLAN to minimize harm to the community. Disadvantaged groups in online social communities [5].

Fair Causality Inference: There have been many studies on the use of causal and graph models while designing fairness algorithms [5].

Fair Representation Learning: *Variational Auto Encoders:* Learning to fairly represent and avoid unfair interference of sensitive attributes has been suggested in many studies [5]. Christos Louizos' Variational Fair autocoding toolkit uses the deletion of sensitive variables information to obtain a reason-able representation [5]. A maximum mean discrepancy regularizer is used to obtain invariance in the posterior distribution over latent variables [5]. Author Elliot Creager et al. proposed a flexible fair representation learning method (by classification that separates information from many sensitive attributes) [5]. This approach addresses the concept of demographic fairness, which can deal for multiple sensitive attributes or any subset thereof.

Fair NLP (Fair Natural Language Processing): *Word embedding*: Tolga Bolukbasi et al. proposed a method to debiased word embeddings by propos-ing a method that respects the embeddings for gender-specific words but de-biased embeddings for gender-neutral words [5]. Besides, other works give new directions and propose methods to eliminate errors while using embed-ding [5].

Stage 2: In-Processing

The *In-processing* stage recognizes that modeling techniques often become biased by dominant features, other distributional effects, or try to find a balance between multiple model objectives. For example, we having a model which is both accurate and fair. The *In-processing* stage tackle this by often incorporating one or more fairness metrics into the model optimization functions in a bid to converge towards a model parameterization that maximizes performance and fairness [14].

At this stage, there are also various techniques, that have been proposed to meet some of the definitions of fairness, such as:

Fair Classification: Classification must ensure fairness and not cause harm to the affected subjects [5]. To ensure that, Quite several methods have been pro-posed [1, 5] that satisfy certain definitions of fairness. Specifically: Subgroup fairness in [5]; Equality of chance and equal odds in [8]; Different treatment and different impact in [11]; Odds are balanced in [5]. And some other methods [5].

Fair Regression: Author Richard Berk et al. have proposed a method of fair-ness regression along with evaluating it by a measure introduced as "value of fairness" (POF - Price Of Fairness) to measure the accuracy-fairness trade-off [5]. They offer three different categories: Individual fairness, group fairness, and hybrid fairness.

Stage 3: Post-processing

The *Post-processing* stage recognizes that the actual output of an ML model may be unfair to one or more protected variables and/or subgroup(s) within the protected variable. Thus, The *Post-processing* stage tends to apply transformations to model output to improve prediction fairness. *Post-processing* is one of the most flexible stages as it only needs access to the predictions and sensitive attribute information, without requiring access to the actual algorithms and ML models. This makes them applicable for black-box scenarios where not the entire ML pipeline is exposed [14].

At this stage, various techniques have been proposed to meet some definitions of Fair-ness in AI, such as:

Structured Prediction: Author Alekh Agarwal et al., fearing that structured predictive models risk taking advantage of social bias, proposed a correction algorithm called RBA (reducing bias amplification) [8]. RBA ensures that the model predictions follow the same distribution in the training data.

Language Model: A study published by Shikha Bordia and Samuel Bowman proposed a technique to measure gender bias in texts [5]. This technique is based on texts generated from a neural network-based language model, iteratively trained on a corpus of text along with measuring biases in the training text itself.

In addition, an approach to minimize bias in AI software through the convex fairness criterion has been proposed by Naman Goel in [1]. In the research directions close to [1], there have also been many studies on making machine learning fair and non-discriminatory such as [5, 8, 11]. Another line of research is preprocessing of training data [12, 16] to enable equitable learning. The most notable of these is the concept-predictability (and fairness) [17] to measure and correct bias in the training data. The theoretical work of Dwork et al. [18] discusses the concept of individual justice to ensure that people who are similar in terms of insensitive characteristics are treated the same way, Hardt et al. discuss an approach to modifying a classifier (post-training) to make its decisions look non-discriminatory [8]. Kleinberg, Mullainathan, and Raghavan studied theoretical trade-offs and the incompatibility of different conceptions of non-dis-crimination in [8]. Some methods to ensure fairness are summarized in Table 2.

Table 2. Summary of some current approaches to ensure fairness

Solutions	Methods	Definition used	Ref.
Pre- processing	Bias Mitiganting	*Group Fairness and individual Fairness are both used to reduce bias*	[6, 8, 17]
	Community detection	*Equal Opportunity*	[6, 8]
	Fair Causality Inference	*Demographic Parity; Fairness Through Unawareness*	[1, 5, 8, 12]
	Variational Auto Encoders	*Demographic Parity*	[5, 8]
	Word embedding	*Demographic Parity; Equal Op portunity*	[5]
In- processing	Fair classification	Group Fairness: *Equal Opportunity; Equalized Odds*	[3, 6, 8]
	Fair Regression	*Demographic Parity*	[11, 18]
Post- processing	Structured prediction	*Fairness Through Awareness*	[5]
	Language Model	*Conditional Statistical Parity*	[5]

3 Challenges for Engineering AI Fairness in Software-Intensive Systems

We have studied the literature based on published surveys of researchers on the issue of fairness in AI as well as studies related to the quality of AI software systems [1, 5, 8, 9, 11, 14, 22–25] to explore the most common challenges to fairness in AI. Based on the assessments of those studies, we have compiled the fairness challenges in AI presented in the following sections. The authors have given many different definitions of fairness in AI software as well as mentioned many approaches to ensure fair-ness in AI software. It can be seen that, although there have been many studies on this issue, there are still many difficulties and challenges for scientists. By reviewing the above literature, we find that the challenges associated with this issue focus on the following four areas:

Area 1: Concepting fairness - *Challenges related to the concept of fairness in software* In the literature [5, 8, 11, 14, 21, 22], the authors gave many different definitions of fairness in AI software. It can be seen that, although there have been many studies on this issue, there seems to be no consensus on the definition of fairness in AI software. In addition, there are actually a lot of fairness criteria proposed but the concepts of AI software fairness need to be quantified appropriately to facilitate the determination and testing of fairness in AI software. On the other hand, fairness is an abstract concept and depends on the subjective views of people and the political system of each country. Therefore, the quantitative definition of the issue of fairness to suit the context is a matter of serious consideration.

Area 2: Attribute-based assessment - *Challenges related to attribute-based assessment of data fairness.* Also according to the above documents, we find that the current definitions of fairness are all trying to optimize the model, but the definitions of fairness do not balance between individual justice and fairness. Group. This is a challenge for those who study fairness in AI.

Area 3: Fairness and performance - *Challenges related to the relationship between fairness and model performance.* In [1, 24] have shown the trade-offs between fair-ness and the performance of the model and also made statistical evaluations on this issue. This indicates, most of the current definitions of fairness negatively affect the performance of the model and are sometimes seen as a trade-off issue.

Area 4: Human factor in AI systems - *Challenges related to the impact of human factors on the success of fair software systems.* In [8], Simon Caton made very clear statistics and analysis about challenges related to the impact of human factors on the success of fair software systems. Recent advances in science and technology have increased the gap in technical understanding compared to traditional models, which places high demands on defining, enhancing, and verifying fairness as well as the de-sign of AI software systems.

4 Conclusions

In this paper, we have mentioned three issues in order to answer the three questions that we raised in the introduction. Firstly, we have compiled the definitions of fairness (especially the quantitative definitions of fairness), that are used by researchers when analyzing the issue of fairness in AI software. The construction, quantification, and expansion of the concepts of fairness in AI software are increasingly interested and developed in current SE researches. Many current types of research are treating fairness as an important quality requirement in AI/ML software. The element of fairness is mentioned in all stages in the software development process, so the research and development of an appropriate fair definition will be a highlight in the field of SE research. Secondly, we have synthesized the approaches to solving the issue of fairness that have been proposed by many researchers, namely, we have outlined three stages applied to approach fairness of AI software: *Pre-processing, In-processing, and Post-processing*. In addition, we also outlined four methods and techniques to approach software fairness: *Mitigating bias; Fair machine learning; Learning about fair representation; Fair natural language processing*. We also specify what fairness definitions are used in those approaches and techniques. Thus, with current SE, the approach to solving the problem of fairness in AI software gives priority to an approach that emphasizes intervention. In particular, building definitions of fairness based on probability assessment techniques to ensure fairness in AI software is considered an appropriate solution. Thirdly, we have summarized the difficulties and challenges as well as suggested solutions to overcome those difficulties in the research on AI Software fairness. Based on these syntheses, in the future, we will learn to come up with a solution to check fairness in AI software by building a suitable definition of fairness as well as finding an approach to ensure fairness without affecting other requirements of the AI software. We hope that our summary will help readers when intending to research-related issues.

References

1. Goel, N., Yaghini, M., Faltings, B.: Non-discriminatory machine learning through convex fairness criteria. In: Proceedings of the 2018 AAAI/ACM Conference on AI, Ethics, and Society, New Orleans LA USA, p. 116. Dec 2018. https://doi.org/10.1145/3278721.3278722
2. Datta, A., Tschantz, M.C., Datta, A.: Automated experiments on ad privacy settings: a tale of opacity, choice, and discrimination. ArXiv14086491 Cs, Mar. 2015. http://arxiv.org/abs/1408.6491. Accessed 08 Apr 2021
3. Statement on Algorithmic Transparency and Accountability. https://iapp.org/resources/article/statement-on-algorithmic-transparency-and-accountability/. Accessed 15 Jun 2021
4. Big Data: A Report on Algorithmic Systems, Opportunity, and Civil Rights. Benton Foundation, May 05, 2016. https://www.benton.org/headlines/big-data-report-algorithmic-systems-opportunity-and-civil-rights. Accessed 15 Jun 2021.
5. Mehrabi, N., Morstatter, F., Saxena, N., Lerman, K., Galstyan, A.: A survey on bias and fairness in machine learning. ArXiv190809635 Cs, Sep. 2019. http://arxiv.org/abs/1908.09635. Accessed 08 Apr 2021
6. Chakraborty, J., Majumder, S., Yu, Z., Menzies, T.: Fairway: A Way to Build Fair ML Software. In: Proceedings of 28th ACM Joint European Software Engineering Conference and Symposium on the Foundations of Software Engineering, pp. 654–665, Nov. 2020. https://doi.org/10.1145/3368089.3409697
7. Markham, A.N., Buchanan, E.: Ethical Decision-Making and Internet Research: Version 2.0 Recommendations from the AoIR Ethics Working Committee. undefined, 2012. https://aoir.org/reports/ethics2.pdf. Accessed 16 Apr 2023
8. Caton, S., Haas, C.: Fairness in machine learning: a survey. ArXiv201004053 Cs Stat, Oct. 2020. http://arxiv.org/abs/2010.04053. Accessed 08 Apr 2021

9. Robert, L.P., Pierce, C., Marquis, L., Kim, S., Alahmad, R.: Designing fair AI for managing employees in organizations: a review, critique, and design agenda. Human–Computer Interact. Mar. 2020. https://www.tandfonline.com/doi/abs/10.1080/07370024.2020.1735391. Accessed: Jun. 16, 2021
10. Heyn, H.-M., et al.: Requirement Engineering Challenges for AI-intense Systems Development. ArXiv210310270 Cs, Mar. 2021. http://arxiv.org/abs/2103.10270. Accessed 07 Jun 2021
11. Hutchinson, B., Mitchell, M.: 50 years of test (Un) fairness: lessons for machine learning. In: Proceedings of Conference on Fairness, Accountability, and Transparency, pp. 49–58, Jan. 2019. https://doi.org/10.1145/3287560.3287600
12. Zhang, T., zhu, T., Li, J., Han, M., Zhou, W., Yu, P.: Fairness in semi-supervised learning: unlabeled data help to reduce discrimination. IEEE Trans. Knowl. Data Eng. **PP**, 1 (2020). https://doi.org/10.1109/TKDE.2020.3002567
13. Kusner, M.J., Loftus, J., Russell, C., Silva, R.: Counterfactual Fairness. Advances in Neural Information Processing Systems. vol. 30, 2017. https://pa-pers.nips.cc/paper/2017/hash/a486cd07e4ac3d270571622f4f316ec5-Abstract.html. Accessed 03 Jun 2021
14. Barocas, S., Hardt, M., Narayanan, A.: Fairness in Machine Learning, p. 181
15. Chouldechova, A.: Fair prediction with disparate impact: A study of bias in recidivism prediction instruments. ArXiv161007524 Cs Stat, Oct. 2016. http://arxiv.org/abs/1610.07524. Accessed 24 May 2021
16. Kamiran, F., Calders, T.: Classifying without discriminating. In: Control and Communication 2009 2nd International Conference on Computer, pp. 1–6. Feb 2009. https://doi.org/10.1109/IC4.2009.4909197
17. Feldman, M., Friedler, S.A., Moeller, J., Scheidegger, C., Venkatasubramanian, S.: Certifying and Removing Disparate Impact. In Proceedings of the 21th ACM SIGKDD International Conference on Knowledge Discovery and Data Mining, New York, USA, pp. 259–268. Aug. 2015. https://doi.org/10.1145/2783258.2783311
18. Alipourfard, N., Fennell, P.G., Lerman, K.: Can you Trust the Trend: Discovering Simpson's Paradoxes in Social Data. ArXiv180104385 Cs, Jan. 2018. http://arxiv.org/abs/1801.04385. Accessed 08 Apr 2021
19. Jones, G.P., Hickey, J.M., Di Stefano, P.G., Dhanjal, C., Stoddart, L.C., Vasileiou, V.: Metrics and methods for a systematic comparison of fairness-aware machine learning algorithms. ArXiv201003986 Cs, Oct. 2020. http://arxiv.org/abs/2010.03986. Accessed 08 Apr 2021.
20. Zhang, J.M., Harman, M., Ma, L., Liu, Y.: Machine Learning Testing: Survey, Land-scapes and Horizons. ArXiv190610742 Cs Stat, Dec. 2019. http://arxiv.org/abs/1906.10742. Accessed 08 Apr 2021
21. Martínez-Fernández, S., et al.: Software Engineering for AI-Based Systems: A Survey. ArXiv210501984 Cs, May 2021. http://arxiv.org/abs/2105.01984. Accessed 06 Aug 2021
22. Friedler, S.A., Scheidegger, C., Venkatasubramanian, S., Choudhary, S., Hamilton, E.P. Roth, D.: A comparative study of fairness-enhancing interventions in machine learning. ArXiv180204422 Cs Stat, Feb. 2018. http://arxiv.org/abs/1802.04422. Accessed 08 Apr 2021
23. Verma, S., Rubin, J.: Fairness definitions explained. In: Proceedings of the International Workshop on Software Fairness, New York, USA, pp. 1–7 May 2018. https://doi.org/10.1145/3194770.3194776
24. Dutta, S., Wei, D., Yueksel, H., Chen, P.-Y., Liu, S., Varshney, K.R.: Is there a trade-off between fairness and accuracy? a perspective using mismatched hypothesis testing. ArXiv191007870 Cs Math Stat, Dec. 2020. http://arxiv.org/abs/1910.07870. Accessed 08 Apr 2021.
25. Dwork, C., Hardt, M., Pitassi, T., Reingold, O., Zemel, R.: Fairness through awareness. In: Proceedings of the 3rd Innovations in Theoretical Computer Science Conference. ACM, pp. 214–226 (2012)

26. Brun, Y., Meliou, A.: Software fairness. In: Proceedings of the 2018 26th ACM Joint Meeting on European Software Engineering Conference and Symposium on the Foundations of Software Engineering, pp. 754–759 (2018). https://doi.org/10.1145/3236024.3264838

An Operational Profile for Normative Multi-agent Systems

Yahia Menassel[1,2](✉), Toufik Marir[1], and Farid Mokhati[1]

[1] Research Laboratory on Computer Science's Complex Systems (RELA(CS)2) Laboratory, B.P. 358 – 04000 Oum El Bouaghi, Algeria
[2] Department of Mathematics and Computer Science, University of Tébessa, Tébessa, Algeria
yahia.menassel@univ-tebessa.dz

Abstract. Software reliability engineering (SRE) focus on engineering techniques for developing and maintaining software systems whose reliability can be quantitatively evaluated. Software operational profile (SOP) development is the first important step in software reliability testing process. Several approaches have proposed methodologies to generate the software operational profile. However, almost all the proposed studies have omitted the specificities of the new software paradigms, despite that these new paradigms are generally characterized by their own concepts, methods and methodologies. Multi-agent systems are one of these software paradigms. It will be more beneficial to propose a specific software operational profile development methodology instead of using a general one. In this paper, we propose a development methodology of a specific kind of multi-agent systems (called, normative multi-agent systems). This methodology is described gradually using a case study.

Keywords: SOP · Reliability testing process · Normative multi-agent systems

1 Introduction

Nowadays and with the development of computer technology, software becomes ubiquitous products. Consequently, ensuring the sustainability of such software is becoming increasingly important. In fact, sustainability is actually an important issue, especially in software engineering [1]. Although the relationship between "quality" and "sustainability" is still debated [2], we think that they share some characteristics such as performance and reliability. Known that the reliability refers to the probability that a system will perform its intended function during a specified period under stated conditions [3], reliable software allows using it for a long period of time. In order to measure software reliability, software reliability testing techniques have been developed. The first and most important step in measurement process is Software Operational Profile (SOP), a way to generate the software reliability test cases. SOP corresponds to a quantitative description of how to use a system from user's viewpoint, which is a set of system operational scenarios and their associated probabilities [4, 5]. SOP provides several benefits such as increase productivity, reliability and making testing faster by allocating resources efficiently in

V. Gupta et al. (Eds.): SSEBIM 2022, LNISO 62, pp. 14–28, 2023.
https://doi.org/10.1007/978-3-031-32436-9_2

relation to use and criticality [6]. Identifying an operational profile becomes relevant to ensure software reliability because it detects failures and the faults causing them [7]. Although several works have been proposed in the literature on multi-agent systems testing, none of them focused on testing their reliability. The major goal of this paper is to propose a methodology to develop an operational profile for normative multi-agent systems (NorMAS).

The rest of this paper is structured as follows: in Section 2, we present literature review related to software operational profile development. Section 3 describes the detailed development process of the operational profile for NorMAS. The proposed methodology is illustrated by a detailed case study. In Section 4, we draw some conclusions and give some future directions.

2 Related Works

Several works related to SOP development can be found in literature. Musa introduced SOP in [4] and [5]. The development process contains five steps namely customer profile, user profile, system mode profile, functional profile and operational profile. He illustrated his approach with a case study of telecommunication switching system. A method devised for creating operational profiles by decomposing an OP into two components: a configuration profile and a usage profile is presented in [8]. This paper provides examples from the project in which this method was first used (AT&T Private Branch Exchange). This approach is most useful when the product's customers, usage, and configurations are varied and complex.

In [9], a structured method for developing an operational profile for interactive users is explained and illustrated with an example. It also shows the use of a test case selection tool that facilitates the tester's task by using as input information obtained while developing the operational profile. Another work proposes a refinement methodology for the generation of more accurate operational profiles that truly represent the diverse customer usage patterns [10]. Clustering analysis supports the refinement methodology for identifying groups of customers with similar characteristics. The operational profile model is extended in [11]. The extended version is composed of a process profile, a structural profile and a data profile in addition to operational profile. Process profile describes the processes and the frequencies with which they occur in a typical user application. Structure profile is the structure of the system on which application is running. The values of the inputs to the application from one or more users present data profile.

A systematic operational profile development for software components is presented in [12]. The method uses both intended usage assumptions and usage data to discover a usage structure, usage distribution and characteristics of parameters. In [13], a software that had undergone normal scheduled testing cycle of the company before is tested through an operational profile. The SOP is developed and test cases are allocated.

A multi-agent framework to automatically generate an operational profile for distributed systems is introduced in [7]. This work also proposes new composed metrics to determine the operations criticality. A framework is proposed in [14] to develop SOP based test case allocation using fuzzy logic. This work presents two case studies to show the applicability of the proposed framework. Another method based on uniform design has been proposed to develop SOP two years later in [15]. A review of a SOP, its extension, enactment, and shortcomings are presented in [6] while [16] is devoted to the survey, analysis, and classification of operational profiles that characterize the type and frequency of software inputs. With some modifications in Musa's approach, a methodology of SOP development in effective way is proposed in [17]. More profiles have been added in the existing model to make efficient operational profile, usage profile and configuration profile. The usage profile shows that users are not same in their usage proportion when the configuration profile defines the way customer set up their system.

Recently, doctoral research is presented in [18] that investigates the use of SOP as a resource to ensure and improve software quality from the point of view of users. The research proposes a strategy in which the users use the SOP to evaluate and adapt an existing test suite and to rank the impact of the defects in the operation of the software, contributing to the pricing of the defects. A behavior-driven development approach [19] is used as an information source for the semi-automated generation of the operational profile. The proposed approach was evaluated on the Diaspora social network software, an open-source project.

According to above presented works, we can conclude two main remarks. Firstly, despite that the first works appeared thirty years ago, this field remains an active field of research with new contributions [17–19]. Secondly, the evolution of the software development paradigms has been omitted in the proposed studies. In fact, each new software paradigm introduces its own concepts, methods and methodologies. Obviously, these concepts will affect the whole development software process. In this paper, we will develop an operational profile for normative multi-agent systems.

3 Developing Normative Multi-agent Systems Operational Profile

According to [5], developing an operational profile involves as many as five steps. The same number of steps makes up the methodology proposed to develop the operational profile of normative multi-agent systems (NorMASOP). As shown in Fig. 1, the steps are as follows: Find the NorMAS customer role profile, Identify the NorMAS user role profile, Define the NorMAS mode profile, Determine the NorMAS goal profile and determine the NorMAS operational profile.

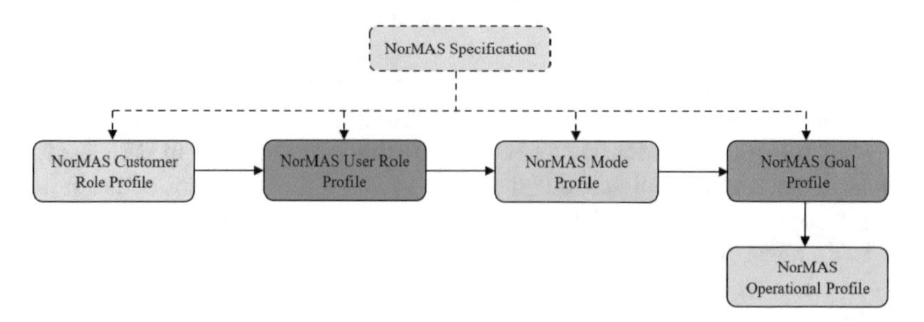

Fig. 1. NorMASOP proposed development methodology.

In order to illustrate our methodology, we consider the "University" system as an example of normative multi-agent systems presented in [20] and shown in Fig. 2. In fact, this case study is a typical example of normative multi-agent systems. It presents almost all the related concepts of such systems, like organizations, roles, agents and norms. Indeed, the specification of this system can be considered as a specification of such system simulation software.

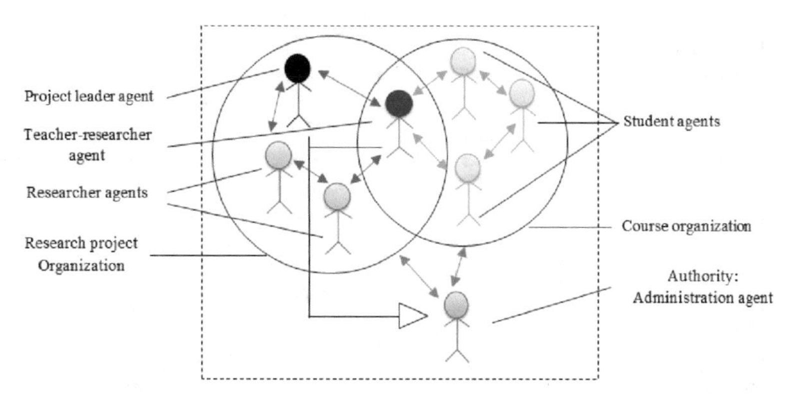

Fig. 2. NorMAS "University".

3.1 NorMAS Specification

The global goal of the "University" system is divided into two sub-goals: research and education. These sub-goals are divided into a set of basic goals that the system agents are committed to satisfy. In turn, the basic goals are grouped into missions and distributed to the various system agents [20]. All agents are obliged, permitted or recommended to satisfy their missions. For example, Teacher-Researcher is obliged to satisfy the mission (Mission-T) defined by the goals teach, evaluate, research and respect regulations. The same agent is allowed to occupy a senior position. Table 1 describes the different norms, which each agent role is assigned to a mission according to norms type (Obligation, Permission, and Recommendation). The administration agent as authority monitors the application of these norms. The administration agent can punish an agent if he does not satisfy an obligation norm, and rewards another if he satisfies a permission or recommendation norm.

3.2 NorMAS Customer Role Profile

Finding the customer role profile is the first step in development process. A customer is an actor, group of actors that will acquire the system. A customer role group is defined by a set of customer roles that will use the system in the same way. The set of customer role groups and their associated occurrence probability defines the customer profile. In case of university system, customer role profile is not used because there is no customer role type. Therefore, the occurrence probability that should be associated with the customer role profile is 1.

Table 1. Normative specification of university system.

Norm	Agent role	Type	Mission	Authority role
Nor1	Teacher- Researcher	Obligation	Mission-T = {goal1(Teach), goal2(Evaluate), goal3(Research), goal4(Respect regulations)}	Administration, Project leader
Nor2	Teacher-Researcher	Permission	Mission-T = {goal5(Occupy a senior position)}	Administration
Nor3	Student	Obligation	Mission-S = {goal6(Guided & Autonomous practice), goal7(Pass evaluation), goal4(Respect regulations)}	Teacher-Researcher, Administration
Nor4	Student	Recommendation	Mission-S = {goal8(Attend courses)}	Teacher- Researcher
Nor5	Researcher	Obligation	Mission-R = {goal9(Collaborate with researchers), goal10(Advance in research), goal11(Attend seminars), goal4(Respect regulations)}	Project leader
Nor6	Researcher	Permission	Mission-R = {goal1(Teach)}	Administration
Nor7	Project leader	Obligation	Mission-Pl = {goal3(Research), goal12(Make industrial contacts), goal13(Sign contracts), goal14(Propose approach), goal15(Supervising researcher), goal4(Respect regulations)}	Administration
Nor8	Project leader	Permission	Mission-Pl = {goal5(Occupy a senior position)}	Administration
Nor9	Administration	Permission	Mission-A = {goal16(Punish), goal17(Recompense)}	-

3.3 NorMAS User Role Profile

A user role profile is not necessarily same as a customer role profile. A user can be an agent or actor, group of agents or actors that uses the system. A user role group is a set of user roles that will use the system in the same way. The user role profile is the user role groups who use the system differently associated with their occurrence probability. The user role profile must be created after defining the customer role profile. If there are multiple customer role groups, then user role groups' probabilities should be multiplied by customer role groups' probabilities to obtain their overall probabilities.

Table 2 shows the user role profile of university system. We identified five types of user role: Teacher-Researcher, Student, Researcher, Project leader and Administration. Also, we assumed that their occurrence probabilities are 0.13, 0.7, 0.06, 0.01 and 0.1 respectively. We note that the overall user role group probabilities are the same as the user role group occurrence probabilities because the customer role groups probabilities are 1 in our case, as it is explained previously.

Table 2. User role profile of university system.

User role	Overall user role group probability
Teacher-Researcher	0.13
Student	0.7
Researcher	0.06
Project leader	0.01
Administration	0.1

3.4 NorMAS Mode Profile

A NorMAS mode is defined as a set of goals and sub-goals grouped to determine execution environment. For each system mode, an operational profile must be determined. A set of system modes and their occurrence probability are called NorMAS mode profile. We can identify two modes in the university system: course mode and research mode. We assume that course mode uses 80% of teacher-researcher role, 100% of student role and 50% of administration role. The rest of the percentages of teacher-researcher and administration (20% and 50%) with 100% of researcher and project leader roles are used by the research mode. The overall probability of course mode is calculated by multiplying the occurrence probabilities of user role groups (teacher-researcher, 0.13; student, 0.7; and administration, 0.1) by the proportion of use (0.8; 1; and 0.5) respectively and adding the results. Using the same formula, we calculate the overall probability of research mode. The derived system mode profile is shown in Table 3.

Table 3. User role profile of university system.

System mode	Occurrence Probability
Course Mode	0.854
Research Mode	0.146

3.5 NorMAS Goal Profile

As its name suggests, the goal profile is a set of goals with their occurrence probability. It must be developed for each system mode and its goals are defined at the design level. To develop the goal profile, it is necessary to find out the initial list of goals that describe the main system goal and environmental variables that can affect system goals during execution. Then, initial goal list and environmental variables are combined to find out final goal list. Due to space limitations, we will explain the last two steps of the proposed methodology through a single mode, which is the course mode.

Initial Goal List. In this step, we need to identify the basic agent goal list of system. We can get it from system specification described in section A. Initial goal list of course mode is shown in Table 4.

Table 4. Initial goal list of course mode.

Goals	Agent role	Course Mode (0.854)	
		Occurrence probability	Overall occurrence probability
Goal1 (Teach)	Teacher-Researcher	0.2	1.708
Goal2 (Evaluate)	Teacher-Researcher	0.1	0.0854
Goal4 (Respect regulations)	Teacher-Researcher, Student	0.1	0.0854
Goal5 (Occupy a senior position)	Teacher-Researcher	0.05	0.0427
Goal6 (Guided & Autonomous practice)	Student	0.2	1.708
Goal7 (Pass evaluation)	Student	0.1	0.0854
Goal8 (Attend courses)	Student	0.1	0.0854
Goal16(Punish)	Administration	0.075	0.06405
Goal17(recompense)	Administration	0.075	0.06405

Table 5. Environmental variables of university system.

NorMAS Goals	Environmental Variables	
	NorMAS agent and actor variables	NorMAS variables
Goal1 (Teach)	Absence of student role (Abs-SR)	NorMAS Resources Problems (NorMAS-RP)
Goal2 (Evaluate)		
Goal3 (Research)	Absence of project leader role (Abs-LR)	
Goal4 (Respect regulations)	–	
Goal5 (Occupy a senior position)		
Goal6 (Guided & Autonomous practice)	Absence of teacher-researcher role (Abs-TR)	
Goal7 (Pass evaluation)		
Goal8 (Attend courses)		
Goal9 (Collaborate with researchers)	–	
Goal10 (Advance in research)	Absence of project leader role (Abs-LR)	
Goal11 (Attend seminars)	–	
Goal12 (Make industrial contacts)		
Goal13 (Sign contracts)		
Goal14 (Propose approach)		
Goal15 (Supervising researcher)	Absence of Researcher role (Abs-RR)	
Goal16(Punish)	Absence of student and teacher-researcher roles (Abs-SR&TR)	
Goal17(recompense)		

Environmental Variables. Environmental variables are the events that can affect the system execution. According to the specificities of multi-agent systems, we need to consider two types of environmental variables: variables related to the system itself and those related to the different system agents and actors. In our case, we consider the environmental variables shown in Table 5. The problem of didactic, pedagogical and material resources (NorMAS-RP) is considered as a variable related to the system. For the different agents of the system, we consider the absence of role variable (absence of student role, absence of project leader role …).

Final Goal List. After considering the environmental factors, we can calculate final goal list by multiplying initial goal list and environment variables. Table 6 shows final goal profile of course mode.

Table 6. Goal profile of course mode.

Goals	Norm type	Agent role	Course Mode (0.854)	
			Environmental variables	Occurrence probability
Goal1 (Teach)	Obligation	Teacher-Researcher	NorMAS-RP (0.01)	0.001708
			Abs-SR (0.01)	0.001708
			Normal conditions (0.98)	0.167384
Goal2 (Evaluate)			NorMAS-RP (0.01)	0.000854
			Abs-SR (0.01)	0.000854
			Normal conditions (0.98)	0.083692
Goal4 (Respect regulations)		Teacher-Researcher, Student	NorMAS-RP (0.01)	0.000854
			Normal conditions (0.99)	0.084546
Goal5 (Occupy a senior position)	Permission	Teacher-Researcher	NorMAS-RP (0.01)	0.000427
			Normal conditions (0.99)	0.042273
Goal6 (Guided & Autonomous practice)	Obligation	Student	NorMAS-RP (0.01)	0.001708
			Abs-TR (0.01)	0.001708
			Normal conditions (0.98)	0.167384
Goal7 (Pass evaluation)	Obligation	Student	NorMAS-RP (0.01)	0.000854
			Abs-TR (0.01)	0.000854

(*continued*)

Table 6. (*continued*)

Goals	Norm type	Agent role	Course Mode (0.854)	
			Environmental variables	Occurrence probability
			Normal conditions (0.98)	0.083692
Goal8 (Attend courses)	Recommendation		NorMAS-RP (0.01)	0.000854
			Abs-TR (0.01)	0.000854
			Normal conditions (0.98)	0.083692
Goal16 (Punish)	Permission	Administration	NorMAS-RP (0.01)	0.0006405
			Abs-SR&TR (0.001)	0.00006405
			Normal conditions (0.989)	0.06334545
Goal17 (recompense)			NorMAS-RP (0.01)	0.0006405
			Abs-SR&TR (0.001)	0.00006405
			Normal conditions (0.989)	0.06334545

3.6 NorMAS Operational Profile

NorMAS Operational profile is a set of operations with their occurrence probability. Each goal from each system mode breaks down into a set of sub-goals, which are in turn broken down into a set of actions (operations). For example, teaching scenario (goal 1), presented by an activity diagram in Fig. 3 includes four actions: Prepare course, during which teacher-researcher prepares his or her presentation; Present course, during which the teacher-researcher demonstrates the learning to be achieved; Guided practice, during which teacher-researcher, with the students, performs tasks similar to those performed during course presentation; and Prepare autonomous practice, during which the teacher-researcher prepares a personal work and lets the student reinvests alone what he has learned during the two previous steps.

The overall occurrence probability of each action is calculated by multiplying the initial action probability by the occurrence probability of its goal. If the goal includes only one action, the overall occurrence probability of that action is the same as the overall occurrence probability of its goal. The operational profile of course mode will be shown in the following Table 7.

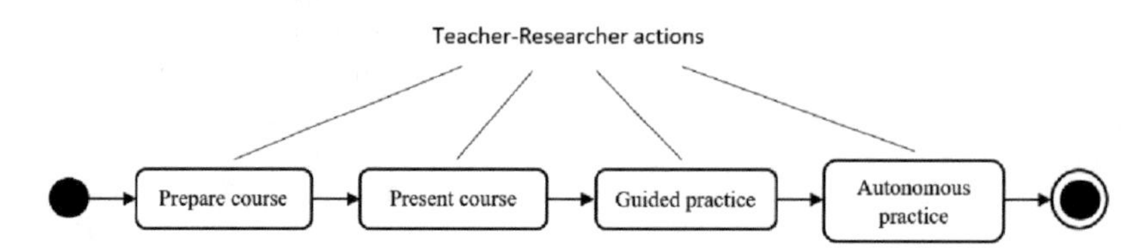

Fig. 3. Teaching goal scenario.

Table 7. Operational profile of course mode.

Agent role	Actions (operations)	Course Mode (0.854)		
		Occurrence probability	Environmental variables	Overall occurrence probability
Teacher-Researcher	Goal1 (Teach)			
	Prepare course	0.2	NorMAS-RP (0.01)	0.0003416
			Abs-SR (0.01)	0.0003416
			Normal conditions (0.98)	0.0334768
	Present course	0.3	NorMAS-RP (0.01)	0.0005124
			Abs-SR (0.01)	0.0005124
			Normal conditions (0.98)	0.0502152
	Guided practice	0.3	NorMAS-RP (0.01)	0.0005124
			Abs-SR (0.01)	0.0005124
			Normal conditions (0.98)	0.0502152
	Prepare autonomous practice	0.2	NorMAS-RP (0.01)	0.0003416
			Abs-SR (0.01)	0.0003416
			Normal conditions (0.98)	0.0334768

(*continued*)

Table 7. (*continued*)

Agent role	Actions (operations)	Course Mode (0.854)		
		Occurrence probability	Environmental variables	Overall occurrence probability
Teacher-Researcher	Goal2 (Evaluate)			
	Prepare the evaluation work	0.2	NorMAS-RP (0.01)	0.0001708
			Abs-SR (0.01)	0.0001708
			Normal conditions (0.98)	0.0167384
	Prepare typical answer	0.2	NorMAS-RP (0.01)	0.0001708
			Abs-SR (0.01)	0.0001708
			Normal conditions (0.98)	0.0167384
	Correct evaluation work	0.5	NorMAS-RP (0.01)	0.000427
			Abs-SR (0.01)	0.000427
			Normal conditions (0.98)	0.041846
	Assign marks	0.1	NorMAS-RP (0.01)	0.0000854
			Abs-SR (0.01)	0.0000854
			Normal conditions (0.98)	0.0083692
Teacher-Researcher, Student	Goal4 (Respect regulations)			
	Know the regulations	0.2	NorMAS-RP (0.01)	0.0001708
			Normal conditions (0.99)	0.0169092
	Practice rights	0.4	NorMAS-RP (0.01)	0.0003416
			Normal conditions (0.99)	0.0338184
	Apply duties	0.4	NorMAS-RP (0.01)	0.0003416
			Normal conditions (0.99)	0.0338184
Teacher-Researcher	Goal5 (Occupy a senior position)			

(*continued*)

Table 7. (*continued*)

Agent role	Actions (operations)	Course Mode (0.854)		
		Occurrence probability	Environmental variables	Overall occurrence probability
	Occupy a senior position	1	NorMAS-RP (0.01)	0.000427
			Normal conditions (0.99)	0.042273
Student	Goal6 (Guided & Autonomous practice)			
	Achieve guided practice	0.6	NorMAS-RP (0.01)	0.0010248
			Abs-TR (0.01)	0.0010248
			Normal conditions (0.98)	0.1004304
	Achieve autonomous practice	0.4	NorMAS-RP (0.01)	0.0006832
			Abs-TR (0.01)	0.0006832
			Normal conditions (0.98)	0.0669536
	Goal7 (Pass evaluation)			
	Responds to evaluation work	1	NorMAS-RP (0.01)	0.000854
			Abs-TR (0.01)	0.000854
			Normal conditions (0.98)	0.083692
	Goal8 (Attend course)			
	Attend course presentation	1	NorMAS-RP (0.01)	0.000854
			Abs-TR (0.01)	0.000854
			Normal conditions (0.98)	0.083692
Administration	Goal16 (Punish)			
	Apply punishment	1	NorMAS-RP (0.01)	0.0006405
			Abs-SR&TR (0.001)	0.00006405
			Normal conditions (0.989)	0.06334545
	Goal17 (recompense)			
	Offer a recompense	1	NorMAS-RP (0.01)	0.0006405
			Abs-SR&TR (0.001)	0.00006405
			Normal conditions (0.989)	0.06334545

4 Conclusions and Future Directions

In order to test the reliability of normative multi-agent systems (NorMAS), this paper introduces a methodology for developing an operational profile for this type of systems. The proposed methodology is illustrated through a concrete case study of normative multi-agent called university system. This methodology takes into account the specificities of normative multi-agent systems, like using the role and goal concepts. Furthermore, it distinguishes between the agent and the system levels. We think that developing reliable software, can improve its sustainability.

As future work, we plan to extend the proposed methodology to normative MAS reliability testing. This work can be done by generating test cases using the operational profile. In addition, it is possible to propose a methodology that help generating the operation profile using the norms identified in specification phase of the software development.

Acknowledgment. This research work was supported by the General Direction of Scientific Research and Technological Development (DGRSDT) of the Algerian Higher Education and Scientific Research Ministry. We would like to thank the DGRSDT for its support in the achievement of this work.

References

1. Wolfram, N., Lago, P., Osborne, F.: Sustainability in software engineering. In: 2017 Sustainable Internet and ICT for Sustainability (SustainIT), pp. 1–7 (2017)
2. Calero, C., Bertoa, M.-F., Moraga, M.-Á.: Sustainability and quality: Icing on the cake. In: RE4SuSy@ RE, vol. 995 (2013)
3. Koziolek, H.: Operational Profiles for Software Reliability. In: Seminar "Dependability Engineering," pp. 1–17. Oldenburg, Germany (2005)
4. Musa, J.-D.: The operational profile in software reliability engineering: an overview. In: Proceedings Third International Symposium on Software Reliability Engineering, Research Triangle Park, NC, USA, 1992, pp. 140–154 (1992). https://doi.org/10.1109/issre.1992.285850
5. Musa, J.-D.: Operational profiles in software-reliability engineering. In: IEEE Software. vol. 10, no. 2, pp. 14–32 (1993). https://doi.org/10.1109/52.199724
6. Muhammad Ali-Shahid, M., Sulaiman, S.: A review of software operational profile in software reliability engineering. In: 2014 8th. Malaysian Software Engineering Conference (MySEC), Langkawi, Malaysia, pp. 1–6 (2014).https://doi.org/10.1109/MySec.2014.6985980
7. Yamany, H.E., Capretz, M.A.: A multi-agent framework for building an automatic operational profile. In: Elleithy, K. (ed.) Advances and Innovations in Systems, Computing Sciences and Software Engineering, pp. 161–166. Springer Netherlands, Dordrecht (2007). https://doi.org/10.1007/978-1-4020-6264-3_29
8. Juhlin, B.-D.: Implementing operational profiles to measure system reliability. In: Proceedings of the 3rd International Symposium on Software Reliability Engineering (ISSRE 1992), Research Triangle Park, NC, USA, pp. 286–295 (1992).https://doi.org/10.1109/issre.1992.285896

9. Pant, H., Franklin, P., Everett, W.: A structured approach to improving software-reliability using operational profiles. In: Proceedings of Annual Reliability and Maintainability Symposium (RAMS), Anaheim, CA, USA, pp. 142–146 (1994).https://doi.org/10.1109/rams.1994.291097
10. Elbaum, S., Narla, S.: A methodology for operational profile refinement. In: Annual Reliability and Maintainability Symposium 2001 Proceedings. International Symposium on Product Quality and Integrity (Cat. No.01CH37179), Philadelphia, PA, USA, pp. 142–149 (2001). https://doi.org/10.1109/rams.2001.902457
11. Gittens, M., Lutfiyya, H., Bauer, M.: An extended operational profile model. In: 15th International Symposium on Software Reliability Engineering, Saint-Malo, France, pp. 314–325 (2004).https://doi.org/10.1109/issre.2004.8
12. Shukla, R., Carrington, D., Strooper, P.-A.: Systematic operational profile development for software components. In: 11th Asia-Pacific Software Engineering Conference, Busan, Korea (South), pp. 528–537 (2004).https://doi.org/10.1109/apsec.2004.95
13. Arora, S., Misra, R.-B.: Software reliability improvement through operational profile testing. In: Annual Reliability and Maintainability Symposium, 2005. Proceedings. Alexandria, VA, USA, pp. 621–627 (2005).https://doi.org/10.1109/rams.2005.1408433
14. Kumar, K.-S., Misra, R.-B.: Software operational profile-based test case allocation using fuzzy logic. Int. J. Autom. Comput. **04**(4), 388–395 (2007). https://doi.org/10.1007/s11633-007-0388-6
15. Fu, J.-P., Lu, M.-Y.: Develop software operational profile with uniform design. In: 2009 IEEE International Conference on Industrial Engineering and Engineering Management, Hong Kong, China, pp. 842–846 (2009). https://doi.org/10.1109/ieem.2009.5372900
16. Smidts, C., Mutha, C., Rodriguez, M., Gerber, M.-J.: Software testing with an operational profile: OP definition. J. ACM Comput. Surv. (CSUR), **46**(3), 1–39 (2014). https://doi.org/10.1145/2518106
17. Amrita, Y.-K.: Development of software operational profile. Int. J. Appl. Eng. Res. **12**(22), 11865–11873 (2017)
18. Cavamura, L.: Operational profile and software testing: aligning user interest and test strategy. In: 2019 12th IEEE Conference on Software Testing, Validation and Verification (ICST), Xi'an, China, 492–494 (2019).https://doi.org/10.1109/icst.2019.00062
19. Fazzolino, R., Rodrigues, G.-N.: Feature-trace: an approach to generate operational profile and to support regression testing from BDD features. In: Proceedings of the XXXIII Brazilian Symposium on Software Engineering (SBES 2019), pp. 332–336 (2019).https://doi.org/10.1145/3350768.3350781
20. Boudhaouia, A., Mazigh, B., Missaoui, E.: A formal specification and verification of normative multi-agent systems by DisCSP. In: 2017 IEEE/ACS 14th International Conference on Computer Systems and Applications (AICCSA), Hammamet, Tunisia, pp. 399–406 (2017).https://doi.org/10.1109/aiccsa.2017.134

Self-repair Measurement in FPGA-Based Partial Reconfigurable Systems

Mohamed Sedik Chebout(✉), Toufik Marir, and Farid Mokhati

Research Laboratory on Computer Science's Complex Systems (RELA(CS)2), University of Oum El Bouaghi, Oum El Bouaghi, Algeria
{chebout.ms,marir.toufik,mokhati.farid}@univ-oeb.dz

Abstract. Intelligent Embedded Systems (IES) represent a new discipline in which Artificial Intelligence (AI) is coupled with Embedded Systems (ESs) in order to create a new self-X-based Embedded Systems generation. Self-X capabilities, like: self-repair, self-awareness, self-adaptation, etc. brought, for an ES, the ability to reason about their external environments and, as a result, adapt their behavior appropriately. Also, self-repair capability defines the fact of identifying and repairing failures in order to avoid the high cost of the ordinary repair process. FPGA (Field Programmable Gate Arrays) provide support for implementing the self-repair capability based on FPGA Partial Reconfiguration (PR) feature. Despite that there are many approaches which proposed self-repair processes, the measurement aspects are often omitted. Indeed, the measurement allows controlling the repair process and comparing the different approaches. In this paper, we discuss the self-repair capability measurement by proposing a common FPGA PR self-repair process followed by a set of hardware (HW) and software (SW) metrics dedicated particularly for FPGA PR-based IES.

Keywords: FPGA · Partial Reconfiguration · Intelligent Embedded Systems · Self-repair · Metrics

1 Introduction

With the increasing complexity of tasks faced by Embedded Systems (ES) in terms of highly dynamic environmental conditions and changing application characteristics, ESs have been augmented with a level of "smartness" or "intelligence" in order to adapt seamlessly their behavior to the environment changes. Also, Intelligent Embedded Systems (IES) are considered as a novel and promising generation of ES. IES are characterized, among others, by self-X capabilities where "X" variously refers to awareness, repair, configuring, optimization, adaptation, and the like.

In [1], the authors define self-repair by the system's ability to maintain some degree of functionality after a failure has occurred. Failures are caused by faults. This latter is caused by human mistakes (i.e., errors). Self-repair capability, for an IES, defines the fact in which the system should identify and repair failures in both software (SW) and hardware (HW) sections without needs to outside boost. System failures cause an

V. Gupta et al. (Eds.): SSEBIM 2022, LNISO 62, pp. 29–38, 2023.
https://doi.org/10.1007/978-3-031-32436-9_3

excessive cost in terms of repairs and maintenance especially with regard to critical ES such as medical or hazardous applications. In order to avoid the high cost of the repair process, ESs have been enhanced with self-repair capabilities. Producing ESs capable of self-repair requires significant changes for both SW and HW components from current design practices [2].

Moreover, the self-repair capability can improve significantly the sustainability. Recently, several studies have focused on this controversial concept in the field of software engineering [3, 4]. Some researchers have distinguished the sustainability in software engineering and the sustainability for software engineering [5]. In this context, the self-repair capability allows using the system for a long time, despite failures and errors without maintenance activity. Consequently, it improves the sustainability for both "in" and "for" software engineering.

Although self-repair is a very important characteristic of an IES, its measurement is not yet studied deeply. By measurement we mean the process of assigning a value to an attribute. Also, measurement makes rational decisions and avoids subjective ones [6]. By using metrics, we can objectively judge the actions to be performed and the obtained results. Consequently, the self-repair process of IES can be evaluated.

According to [7], Field Programmable Gate Array (FPGA) is a prefabricated silicon device that is customized and can be optimized to cope with specific requirements of the target HW platform. FPGA provides an alternative promising way to implement self-repair capability based on FPGA Partial Reconfiguration (PR) feature. FPGA PR consists of modifying the functional configuration of the device during operation. This study is devoted to addressing the self-repair capability measurement in FPGA-based IES. The major contribution of this paper is twofold: introducing a common self-repair process for IES and proposing the mathematical measurement model by defining a set of SW and HW metrics.

The remainder of this paper is organized as follows: Section 2 presents a literature review of self-repair measurements in embedded systems. An overview of the proposed FPGA PR-based self-repair process is introduced in Sect. 3. The self-repair measurement is outlined in Sect. 4. In Sect. 5, we discuss the paper findings. Finally, we conclude the paper and propose our future work in Sect. 6.

2 State of the Art

When reviewing the literature, self-repair capability of an ES has been handled from different points of view [2, 8] in which self-repair methodology for embedded systems was introduced and designed using UML and genetic algorithms. The authors adopt the self-repair classification of [9] into attributive and functional repair and demonstrate that the feasibility of implementing a self-repairing strategy in embedded systems as attributive repair can initially be used to exploit any inherent redundancy. In addition, the functional repair can be employed to restore the system with reduced functionality when the fault is more substantial [2]. In [10], a survey on self-healing systems was provided. Also, self-healing systems' fundamental principles have been identified based on autonomic computing and self-adapting system research. The survey gives an insight into a major selection of current and past self-healing approaches. In [1], the authors discuss the possibility to shift in approach from specific repair strategies to autonomous self-repair. Moreover, the focus is instead on what can be done to correct a failure that will invariably occur at some point during system execution rather than focusing on reducing the probability of failure. The authors propose a process for self-correcting systems in which the cause of the fault should be underlined first. After that, the steps of fault detection, fault diagnosis and corrective action selection have been identified.

Although these works have considerably forwarded the self-repair capability for embedded systems, they did not take into account the measurement issue of that feature. Most measurement works dedicated to embedded systems focus, mainly, on a very specific metrics like: performance of FPGA-based partial reconfiguration [11], reconfiguration time overhead [12], Throughput Boosting by over-clocking the Partial Reconfiguration [13] and FPGA-based architecture for performance acceleration [14].

We think that proposing metrics of self-repair characteristic is very important since it allows, on one hand, the evaluation of performed actions by the system during the self-repair process and assess their reliability and effectiveness on the other hand. Moreover, these metrics allow comparing different self-repair approaches and processes. In this work, we propose a self-repair process that is supported by mathematical measurement model. The proposed model can be used to evaluate each activity in the proposed process.

3 Partial Reconfiguration Based-Self-repair Process

Partial Reconfiguration (PR) is a kind of FPGA-Runtime Reconfiguration (RTR) which consists of modifying the functional configuration of the device during operation through either HW or SW changes. Also, one or more regions can be reconfigured while the remainder of the FPGA continues to operate in the system. The configuration data produced to program a reconfigurable device is called a bitstream. A partial bitstream is produced when only a portion of a FPGA-based device is to be reconfigured. In this paper, we adopt PR approach for implementing self-repair process. Otherwise stated, the FPGA keep operates when a PR-based self-repair process goes on.

In the best of our knowledge, a self-repair process incorporates four main steps: self-monitoring, self-diagnosing, self-configuration and self-optimization. We propose a common self-repair process (fig. 1) in which a set of self-repair related concepts will be clarified. Self-monitoring consists of detecting and identifying failures. In the context of this work, we suggest the use of Aspect-Oriented Programming (AOP)-based solutions for implementing self-monitoring activity. AspectC/AspectC++ [15, 16] are the practical aspect-oriented extension for C programming language since the major ES are implemented using SystemC libraries. In contrast, self-diagnosing activity will take place in order to locate the source of failure and deciding which corrective action should be selected. Several mechanisms are targeted to implement self-diagnosing activity like: genetic algorithms, case-based reasoning, deep learning-based prediction, etc. Failures detected will be substituted after invoking the self-reconfiguring activity (i.e., loading the selected corrective action (i.e., partial bitstream file)). Finally, in self-optimizing phase, a performance enhancement process will take place by eliminating sub behaviors caused by the previous steps. The corrective action base consists of a set of partial predefined bitstream files. Each partial bitstream file implements an alternative (or a set of) solution of a given failure.

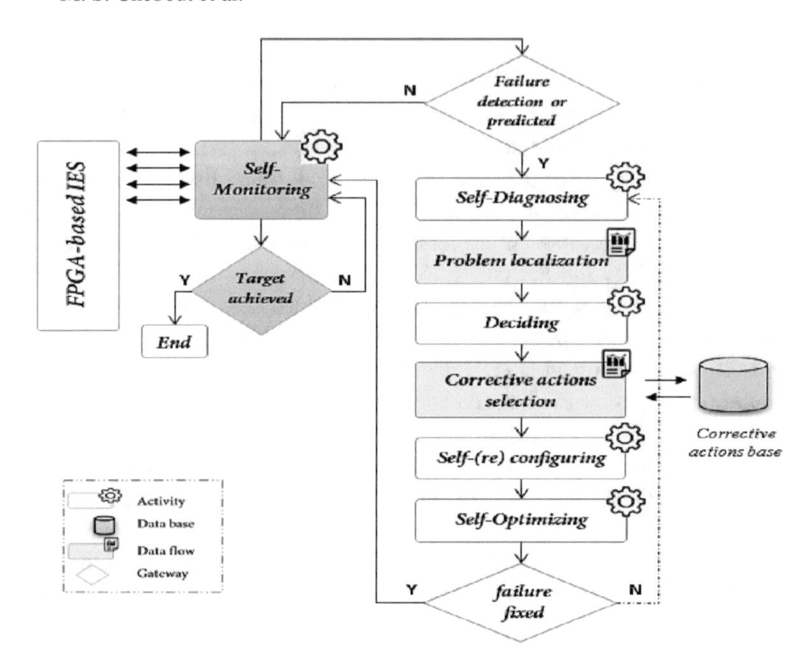

Fig. 1. Self-repair process.

In order to avoid failures, we propose to implement several solutions (bitstream files) for a given failure in order to make more flexibility for the self-repair process. By flexibility, we mean the fact in which a self-repair process will be able to select one or more solutions for a given failure according to its actual situation to satisfy its objectives.

4 Self-repair Measurement

Measurement of given self-X capability for an FPGA-based IES is a function F_{IES}(C) that takes as input a discrete element (C) from a set of capabilities and outputs the value of that capability. F_{IES}(C) is defined as follows:

$$F_{IES}\left(C \in \left\{ \begin{array}{c} SR, SAw, SMa, SCtrl, CA, \\ SAd, SP, SO, SD\, SMo, SCnfg \end{array} \right\}\right) = y \text{ where } y \in [0, +\infty[\quad (1)$$

where SR, SAw, SMa, SCtrl, CA, SAd, SP, SO, SD, SMo and SCnfg stand for self-repair, self-awareness, self-maintenance, self-control, context-aware, self-adaptiveness, self-protecting, self-optimizing, self-diagnosing, self-monitoring and self-configuring respectively. In the case of self-repair capability, we propose to measure it in terms its four activities: self-monitoring (SMo), self-diagnosing (SD), self-configuring (SCnfg) and self-optimizing (SO), as follows:

$$F_{IES}(C = SR) = \frac{\propto_{SMo} F(SMo) + \propto_{SD} F(SD) + \propto_{SCnfg} F(SCnfg) + \propto_{SO} F(SO)}{\propto_{SMo} + \propto_{SD} + \propto_{SCnfg} + \propto_{SO}} \quad (2)$$

knowing that the coefficients $\propto_{SMo}$, $\propto_{SD}$, $\propto_{SCnfg}$ *and* $\propto_{SO}$ represent the importance of the activity. In what follows, we introduce the mathematical model for both SW and HW metrics for each self-repair related activities.

4.1 Self-Monitoring *(SMo)*

In order to measure the self-monitoring activity, we propose several metrics. In fact, some metrics are based on the confusion matrix (CM) that represents the possible configurations for a given fault (Table 1) where TP means that the monitor predicts a failure and it is true, TN means that the monitor predicts that there is no failure and it is true, FP means that the monitor predicts a failure and it is false and FN means that the monitor predicts that there is no failure and it is false.

Table 1. Self-monitoring confusion matrix.

		Actual classes	
		Positive	Negative
Predicted classes	*Positive*	TP	FP
	Negative	FN	TN

Based on the confusion matrix, three metrics are defined: precision, recall and accuracy. We propose, in addition, two more metrics which are frequency of self-monitoring and number of perceived events:

- Precision of self-monitoring (PrecSMo) : this metric refers to the percentage of positive results which are relevant. It is calculated using the confusion matrix as it is presented in the Eq. 2. approximation of this ratio to the 100% rate denotes high accuracy of monitor system.

$$PrecSMo = \frac{\#positives\,predicted\,correctly}{\#positive\,predictions} = \frac{TP}{TP + FP} \tag{3}$$

- Recall of self-monitoring (RecSMo): as it is presented in Eq. 4, this metric refers to the percentage of positive cases correctly classified.

$$RecSMo = \frac{\#positives\,predicted\,correctly}{\#positives\,cases} = \frac{TP}{TP + FN} \tag{4}$$

- Accuracy of self-monitoring (AccSMo): is the percent of predictions that are correct (Eq. 5).

$$AccSMo = \frac{TP + TN}{total} = \frac{TP + TN}{TP + TN + FP + FN} \tag{5}$$

- Frequency of self-monitoring (FSMo): this metric consists of the number of observations made by the monitor in a unit of time (Eq. 6). Obviously, this metric gives information about the frequency of possible failures during the execution of the system.

$$\boldsymbol{FSMo} = \frac{\boldsymbol{\#observations}}{\Delta \boldsymbol{t}} \tag{6}$$

- Number of Perceived Events (NPE): this metric is calculated by the number of events that the monitor can perceive. It reflects the monitoring capability because when the monitor is able to perceive a lot of events, it will be able to detect more failures.

4.2 Self-Diagnosing *(SD)*

For measuring self-diagnosing capability, we propose two metrics:

- The ratio of possible events per possible failures (PEpPF): this metric represents the number of possible events monitored by the system compared by the number of possible failures (Eq. 7). If this ratio is greater than or equal to one (1), it means that there is a strong correlation between events and failures (i. e. the system can easily decide the fault using the captured events). Otherwise, several failures are linked to the same event which complicates the diagnostic activity.

$$\boldsymbol{PEpPF} = \frac{\boldsymbol{\#Events}}{\boldsymbol{\#Failures}} \tag{7}$$

- The ratio of accesses to the corrective actions base (RACBA): this metric defines the fact in which a failure has been located and the corresponding corrective action should be selected. Consequently, this metric is calculated as a ratio of the corresponding corrective action selected (CAS) compared by the total number of located failures (LF) (Eq. 8).

$$\boldsymbol{RACBA} = \frac{\boldsymbol{\#CAS}}{\boldsymbol{\#LF}} \tag{8}$$

4.3 Self-reConfiguration *(SCnfg)*

As we mention above, self-configuration consists of loading one or several selected corrective actions (implemented as bitstream file). Hence, we propose two metrics to measure this activity: Bitstream Loading Time (BLT) and the Number of Corrective Actions (NCL):

- Bitstream Loading Time (BLT): this metric measures the time spent in the loading process. BLT can be measured using profiling based tools like AspectC/AspectC++ profiling techniques like the work presented in [17] or dedicated HW/SW profilers.
- Number of Corrective Actions (NCL): the NCL quantifies the number of corrective actions selected to repair a given failure. It seems obvious that considering the same failure, choosing a minimum of corrective actions has a positive impact on the efficiency of the reconfiguration activity.

Basically, correctness and defectiveness of selected corrective action should be justified. By correctness we mean the ability of the partial bitstream to perform the exact task, as defined in the specification. The defectiveness, in contrast, defines the opposite meaning. In this work, we leave the measurement of correctness and defectiveness capabilities considering that partial bistream files are free from errors.

4.4 Self-Optimizing *(SO)*

To compute SO capability, we propose to calculate the Ratio of Self-Optimizing (RSO). This metric quantifies the number of suboptimal behaviours (SOB) compared to the improved ones (i.e., Improved Behaviours (IB)).

$$\boldsymbol{RSO} = \frac{\#\boldsymbol{SOB}}{\#\boldsymbol{IB}} \tag{9}$$

4.5 Self-repair Process Related Metrics

In addition to metrics proposed for each activity, we also propose some metrics which are related to the whole self-repair process. These metrics are:

- Self-repair effectiveness (SREfctv): effectiveness is the capability of producing a desired results. Also, self-repair effectiveness is defined as the degree to which the PR (i.e. selected partial bitstream) is successful in producing a desired results. For a given failure, we need to compute the Ratio of Reached Repairs RRR compared to the total number of triggered ones (TG) (Eq. 10).

$$\mathbf{SREfctv} = \frac{\#\boldsymbol{RRR}}{\#\boldsymbol{TG}} \tag{10}$$

- PRThroughPut (PRT): the throughput of Partial Reconfiguration is computed off-line by dividing the Bitstream Size (BS) by the Reconfiguration Latency (RL) according to [13]. This metric is calculated according to the Eq. 11.

$$\boldsymbol{PRT} = \frac{\boldsymbol{BS}}{\boldsymbol{RL}} \tag{11}$$

- Self-repair efficiency (SREfcy) : for a given failure, we need to calculate the number of computational resources used by the selected partial bistream. To maximize the SREfcy, we wish to minimize resource usage in terms of time and space. Regarding space, there are up to four aspects of memory usage to consider: the amount of memory needed to hold the code for the partial bitstream, the amount of memory needed for the input data, the amount of memory needed for any output data and the amount of memory needed as working space (i.e., cache memory). Hence, the size of the local memory of the processor can affect significantly PR time.

- Partial Reconfiguration Time (PRT): it can be expressed by the sum of the times spent in each phase of the reconfiguration activity. This last depends strongly on the type of used FPGA. In the case of Virtex-II Pro FPGA, the processor calls the routines to transfer the configuration data from the storage means (SM) like compact flash to its local memory, then to the Internal Configuration (IC) access port configuration cache, and then to the configuration memory (CM).

$$\boldsymbol{PRT = SM_{calls} \times SM_{time} + IC_{calls} \times IC_{time} + CM_{calls} \times CM_{time}} \tag{12}$$

If for every phase, the number of executed processor calls, the amount of configuration data per call, and the time per call are known, we will be able to compute the aggregate reconfiguration time per phase, and finally the total partial reconfiguration time [11].

- Power consumption of PR (PcPR): it defines the amount of power consumed by the PR.
- Speedup: in order to find the speedup between the old FPGA PR and the alternative one, we apply Amdahl's law in a straightforward manner. The speedup can be defined for two different types of quantities: latency and throughput. In the case of latency, the speedup can be calculated according to the Eq. 13 where L1 is the latency of the old PR and L2 is the latency of the alternative PR. The latency can be calculated according to the Eq. 14, where v is the execution speed of the partial bitstream, T is the execution time of the partial bitstream and W is the execution workload of the partial bitstream.

$$\boldsymbol{S_{Latency} = \frac{L1}{L2} = \frac{T_1 W_2}{T_2 W_1}} \tag{13}$$

$$\boldsymbol{L = \frac{1}{v} = \frac{T}{W}} \tag{14}$$

In the case of throughput, the speedup defines the significance of the improvement made by the selected partial bitstream compared to old one. According to Amdahl's law speedup in throughput is calculated as follows:

$$\boldsymbol{S_{throughput} = \frac{Time_{oldPR}}{Time_{newPR}}} \tag{15}$$

5 Discussion

Self-repair capability is an important capability of intelligent embedded systems. It improves both sustainability *for* and *in* software engineering. According to Penzenstadler, B, the sustainability *for* software engineering is related to "the time span is various generations, the function is a satisfaction of needs, and the system is humanity in its ecosystem" [5]. Obviously, developing self-repair systems allow improving the users' satisfaction and decreasing the time developing. In particular, during the maintenance period, the systems can be self-repaired without human intervention to improve the quality of the developed system and increase the life-time of the product. Sustainability *in* software engineering is about developing sustainable software. In this case, using the software must ensure the sustainability. Consequently, self-repair capability improves this concept because repairing the system allows rational use of resources on both SW and HW sides.

6 Conclusion and Future Works

In this paper, we introduce the self-repair measurement for FPGA-based IES by proposing a set of HW and SW metrics. An FPGA PR-based self-repair process has been proposed in order to provides support for HW and SW measurement. The proposed self-repair process is composed of four common activities (self-diagnosing, self-reorganizing, self-monitoring and self-optimizing). Each one of these activities is measured by a set of metrics in which we can measure objectively the self-repair capability. In addition, these metrics can be applied in any operational framework to define the sustainability concept for and in software engineering given the close relationship between this concept and the self-repair capability.

Also, it should be noted that PR depends strongly on the type of used FPGA. In future works, we plan to implement the self-repair process and related metrics on a well-determined FPGA. In addition, we will investigate, deeply, the relationship between the self-repair capability and the sustainability concept in order to propose operational framework to define this novel concept.

Acknowledgment. This research work was supported by the General Direction of Scientific Research and Technological Development (DGRSDT) of the Algerian Higher Education and Scientific Research Ministry. We would like to thank the DGRSDT for its support in the achievement of this work.

References

1. Bell, C., McWilliam, R., Purvis, A., Tiwari, A.: Concepts of self-repairing systems. Measur. Control (United Kingdom) **46**, 176–179 (2013). https://doi.org/10.1177/0020294013492285
2. Coyle, E.A., Maguire, L.P., McGinnity, T.M.: Self-repair of embedded systems. Eng. Appl. Artif. Intell. (2004). https://doi.org/10.1016/j.engappai.2003.11.009
3. Wolfram, N., Lago, P., Osborne, F.: Sustainability in software engineering (2018). https://doi.org/10.23919/SustainIT.2017.8379798.
4. Penzenstadler, B., Bauer, V., Calero, C., Franch, X.: Sustainability in software engineering: a systematic literature review. IET Seminar Digest **2012**(1) (2012). https://doi.org/10.1049/ic.2012.0004
5. Penzenstadler, B.: Towards a definition of sustainability in and for software engineering (2013). https://doi.org/10.1145/2480362.2480585
6. Benaboud, R., Marir, T.: Flexibility measurement model of multi-agent systems. Multiagent Grid Syst. **16**, 309–341 (2020). https://doi.org/10.3233/MGS-200334

7. Yang, H., Zhang, J., Sun, J., Yu, L.: Review of advanced FPGA architectures and technologies. J. Electron. (China) **31**(5), 371–393 (2014). https://doi.org/10.1007/s11767-014-4090-x
8. Coyle, E.A., Maguire, L.P., McGinnity, T.M.: Design philosophy for self-repair of electronic systems using the UML. IEE Proc. Softw. (2002). https://doi.org/10.1049/ip-sen:20020793
9. Umeda, Y., Tomiyama, T., Yoshikawa, H.: A design methodology for self-maintenance machines. J. Mech. Des. Trans. ASME (1995). https://doi.org/10.1115/1.2826688
10. Psaier, H., Dustdar, S.: A survey on self-healing systems: approaches and systems,. Computing (Vienna/New York) (2011). https://doi.org/10.1007/s00607-010-0107-y.
11. Papadimitriou, K., Dollas, A., Hauck, S.: Performance of partial reconfiguration in FPGAsystems. ACM Trans. Reconfigurable Technol. Syst. (2011). https://doi.org/10.1145/2068716.2068722
12. Liu, M., Lu, Z., Kuehn, W., Jantsch, A.: Reducing fpga reconfiguration time overhead using virtual configurations (2010)
13. Nannarelli, A., et al.: Robust throughput boosting for low latency dynamic partial reconfiguration (2017). https://doi.org/10.1109/SOCC.2017.8226013.
14. Diniz, W.F.S., Fremont, V., Fantoni, I., Nóbrega, E.G.O.: An FPGA-based architecture for embedded systems performance acceleration applied to Optimum-Path Forest classifier. Microprocess. Microsyst. **52**, 261–271 (2017). https://doi.org/10.1016/j.micpro.2017.06.013
15. Coady, Y., Kiczales, G., Feeley, M., Smolyn, G.: Using aspectC to improve the modularity of path-specific customization in operating system code (2001). https://doi.org/10.1145/503222.503223.
16. Spinczyk, O., Gal, A., Schröder-Preikschat, W.: {AspectC++}: an aspect-oriented extension to the {C++} programming language (2002)
17. Chebout, M.S., Mokhati, F., Badri, M.: Assessing the effect of aspect refactoring on multi-agent applications. Int. J. Agent Technol. Syst. **7**, 45–66 (2016). https://doi.org/10.4018/ijats.2015070103

Gender Equality in Software Engineering Education – A Study of Female Participation in Customer-Driven Projects

Anh Nguyen-Duc[1](✉) and Letizia Jaccheri[2]

[1] University of South Eastern Norway, Bø i Telemark, Norway
angu@usn.no
[2] Norwegian University of Science and Technology, Trondheim, Norway
letizia.jaccheri@ntnu.no

Abstract. Gender equality, as a part of SDGs, is gaining research attention due to the desire to promote female participation in the engineering sector. The objective of this work is to enhance the understanding of female students' participation in software engineering projects to support gender-aware course optimization. Since 2015, we have investigated the activity profiles of female students in terms of software engineering activities in a fourth-year software project course. Empirical evidence has been collected through surveys, structured interviews and project reports from 39 projects. We found that the active activity areas of female students are project management and requirement engineering, while the areas lacking active involvement are architecture and Scrum methodology. While the findings differ from those of some previous studies, they suggest which course and project settings will facilitate the active participation of females in such project courses.

Keywords: Female participation · Gender · Software engineering project course · Longitudinal study · Mixed approach research

1 Introduction

Software Engineering Education (SEE) is always in a need of evolving to match the demand for practical yet sustainable education. As an important part of SEE, project courses are often a bridge between theories and practices students can gain in an academic context. The main goal of these courses is to equip students with real-life experience and ultimately reduce the gap between education and industrial demands [1, 2]. Recent SEE research has considered how to improve the learning outcome for students in project-based courses [3–6]. From a pedagogic perspective, balancing different stakeholders in the course and maximizing the amount of learning at both the individual and project levels is a non-trivial issue.

Among the 17 Sustainable Development Goals established by the United Nations in 2015, Gender equality is an important goal that closely relates to the academic context. Engineering is a male-dominated field, both in terms of the gender majority and the way in which engineering tasks are framed and valued [7]. A survey in 2016 showed

V. Gupta et al. (Eds.): SSEBIM 2022, LNISO 62, pp. 39–49, 2023.
https://doi.org/10.1007/978-3-031-32436-9_4

that an overwhelming majority (83.3%) of ICT specialists employed in the EU were men [22]. In academia, in the whole of Europe, women take less than 15% of the full professor positions. In the context of project-based courses, female students might be less exposed to varied learning experiences compared to their male peers. For example, female students could be assigned to traditional, nontechnical female roles, such as organizers, secretaries, and writers, whereas men tend to be assigned to more technical roles [7, 9], although both types of roles are important for project success.

Understanding female participation in SEE contributes to improving female students' learning experience during their studies. Appropriate educational approaches would encourage female students' participation in SE as well as their competence and career prospects. Currently, SEE literature is scarce on female activity throughout a software development project. One reason for the lack of investigation of this issue is that female students are often less involved in project work or participate in a modest manner, which makes their learning difficulties to observe. The way in which individuals learn, both in terms of technical competence and non-technical skills, remains an open question.

We conducted an investigation within our course Customer-driven project TDT4290 [21]. To make the study of gender possible, we adjusted the course settings to facilitate the study of female participation in our course. We assigned some female students to each project team and promoted them to active positions as team leaders or managers, although they could still participate in technical tasks. We arranged regular supervision to assist the students with the challenges or issues they encountered. At the end of the course, we conducted surveys and interviews to collect the students' opinions on perceived teamwork, team performance, and learning outcomes. The students also wrote reflection reports as part of the final delivery. In this paper, we address our Research Question (RQ):

RQ: How do female students participate in SE tasks in software projects?

This work has inspired ongoing work in the EUGAIN project[1].The EUGAIN project aims at improving gender balance in Informatics at all levels through the creation of a European network of colleagues working on the forefront of the efforts for gender balance in Informatics in their countries and research communities [23]. At NTNU we also run the IDUN project[2], that proposes a framework to inspire female researchers in their path from PhD to professor.

The paper is organized as follows. Section 2 presents the related work on female participation in software development. Section 3 presents the course setting of the study. Section 4 details the research methodology. Section 5 presents the findings related to the RQ. Section 6 discusses the results and concludes the paper.

2 Female Participation in Software Development

Gender stereotypes - the beliefs that influence expectations about women and men and how they ought to behave—are evident in many social and professional communities. Even though there has been extensive research on gender issues in the social science

[1] https://eugain.eu/.

[2] https://www.ntnu.edu/idun.

field as well as in organizational and management studies [9, 10], the understanding of gender issues in engineering, particularly in SE, is limited [11–15]. Gramß et al. investigated the performance of female students in requirement engineering tasks and showed that less success is expected of females, and their achievements are less likely to be attributed to their personal competence [11]. Technical competencies, such as programming and software architecture, are perceived as being fundamentally male. In terms of coding, there are differences in compatibility and communication levels between same-gender pairs and mixed-gender pairs [12]. Females often develop deficient or inappropriate strategies for complex problem-solving. In code development, females tend to use bottom-up strategies while males are more risk-prone and use more top-down strategies [13].

Regarding quality assurance and testing, female developers expressed lower levels of self-efficacy than males with regard to their debugging abilities [14]. Further, women were less likely than males to accept new debugging features. Although there is no evidence of gender differences in fixing seeded bugs, female developers have been found to introduce more new bugs - which remained unfixed. There are also differences between men and women in terms of the features of programming environments that they use and explore [15]. A study characterizing collaborative learning environments for female participation in software projects described four main concerns: working with others, productivity, confidence, and interest in IT careers [16]. One key finding was that collaboration, emerging from face-to-face meetings, helps female students to build confidence via higher quality products and to reduce the amount of time spent on assignments [16].

3 The Course Setting

The goal of TDT4290 is to teach students "software engineering skills in the context of a development project to make a realistic prototype of an Information System (IS) 'on contract' for a real-world customer" [17]. In the project course, the students are expected to experience the practices of several SE knowledge areas [18], such as Software Requirements, Architecture, Coding, Testing, Project Management, and the Software Engineering Process. In general, students are instructed to adopt Scrum as their way of working.

As a part of the learning objective, the students should be able to understand the importance of good teamwork and team dynamics. Situations that the students are expected to face include handling difficult customers, coordinating team efforts, task distribution, and responsibilities, and collective problem-solving. The students will thus obtain skills related to team communication, task management, collective decision-making, team retrospection, and leadership.

The course lasts for 14 weeks. For a project group of 6–8 students, the available effort per group will lie between 2016 and 2688 person-hours. Groups are randomly assigned. Only one intervention is that we assigned female students in each group and ask them to lead their groups. Each group is assigned a supervisor, who assists the team throughout the courses. A regular meeting is held with customers, which is focused on team dynamics, facilities for project execution, and concerns about communicating with customers. We have a wide range of customers from academia, public sectors, and industries. We selected projects that are suitable to the students in terms of scope, complexity, and accessibility. Examples of projects are:

- Video management software for multiple display devices
- Modules for automated validation of insurance claims
- Mental training mobile apps
- Travel planning software for cyclists
- Machine learning in the detection of abnormal system usage.

4 Research Methodology

The study design was completed in Spring 2015. The data collection was performed annually from Autumn 2015 to Spring 2018. The unit of analysis is individual students. We studied both female students in leading positions and those working as ordinary team members. Various instruments were designed to collect data about student team activities, including project plans, final reports, supervision meeting notes, interviews with team leaders, and a team reflection survey. In total, 39 student teams were studied in depth using a mixed-research approach. Table 1 describes the gender distribution of the students in the courses.

Table 1. Number of students in the course

Year	# Teams	#Male students	#Female students
2015	12	69 (87.4%)	10 (12.6%)
2016	13	72 (83.7%)	14 (16.3%)
2017	14	73 (79.4%)	19 (20.6%)
Total	**39**	**214 (83.3%)**	**43(16.7%)**

4.1 Data Collection Instruments

Data about group's progress is collected from various approaches. We gathered all final reports from student groups, which show how students worked, customer interaction, role fulfillment, what they have learned, and what they could have done differently. At the end of the course, we collect students' opinions on their team dynamics through a survey. The survey is given to each student individually in their final supervision meeting. Besides meta-information about their team and their role within the team, the students have to answer 18 five-point Likert-scale questions about their (1) teams' retrospective meeting practices, (2) collective decision-making, (3) team leadership, and (4) task management practices (the survey instrument is available online[3]). The responses range

[3] https://docs.google.com/document/d/1fkxI4xT5msWf_J6k5ibsOKJNlAJ-Rj3slOIVw8zwJVs.

from "Strongly Agree" to "Strongly Disagree." Each question has a N/A (not applicable) option, which the students can select if they are unsure about or did not experience a particular situation. The survey has been used previously to investigate participants' perceptions of team dynamics [19]. We also asked the leaders of each team to complete a Fiedler least-preferred co-worker (LPC) questionnaire to identify the leadership style. The survey responses were collected in the first three weeks of the project. We were able to gather 33 valid responses from 24 female leaders and 9 male leaders. The full list of responses is made online[4].

4.2 Data Analysis

Based on the two major learning objectives of the course, we aim at characterizing female students' contribution to (1) software engineering (SE) activity and (2) group dynamics. Areas of SE is defined basing on SWEBOK. We focused particularly on female students and derived several measures to capture their activity profiles. Regarding SE activities, we counted the number of female students who actively participated in SE tasks, sum f_i, the percentage of female students participating in the tasks, x_i, and the probability that SE tasks would be led by female students, $\mathrm{pf_i}$. The metrics were as follows:

$$x_i = \frac{\sum f_i}{F} \tag{1}$$

$$pf_i = \frac{\sum f_i}{\sum f_i + \sum m_i} \tag{2}$$

$\mathrm{x_i}$: the portion of female who leads in task i
$\mathrm{f_i}$: a female student that leads task i
$\mathrm{m_i}$: a male student that leads task i
F: the total number of female students

We also collected female students' opinions regarding their influences on each SE knowledge area from the surveys.

Regarding qualitative data, we read all the students' reports and extracted the relevant texts describing individual students' roles and activities in their projects. Structured interviews with students were briefly extracted. For each project, we collected texts about what each student does, what challenges they faced, and then grouped the texts into female and male groups.

With regard to quantitative data, we used IBM SPSS Statistics for descriptive analysis. The data were pre-possessed to remove invalid data points. Besides mean and median values, we also reported standard deviation values to see the variation among the students' responses. We used Wilcoxon–Mann–Whitney (WMW) tests instead of a t-test, which requires normal distribution of the data, to analyze the statistically significant differences among male and female students.

[4] https://tinyurl.com/tdt4290surveyresponses.

5 How do Female Students Participate in Software Engineering Tasks in Software Projects?

Figure 1 presents the distribution of leadership roles in the SE tasks among female students. The effort distribution of female participation is skewed toward non-code activities.

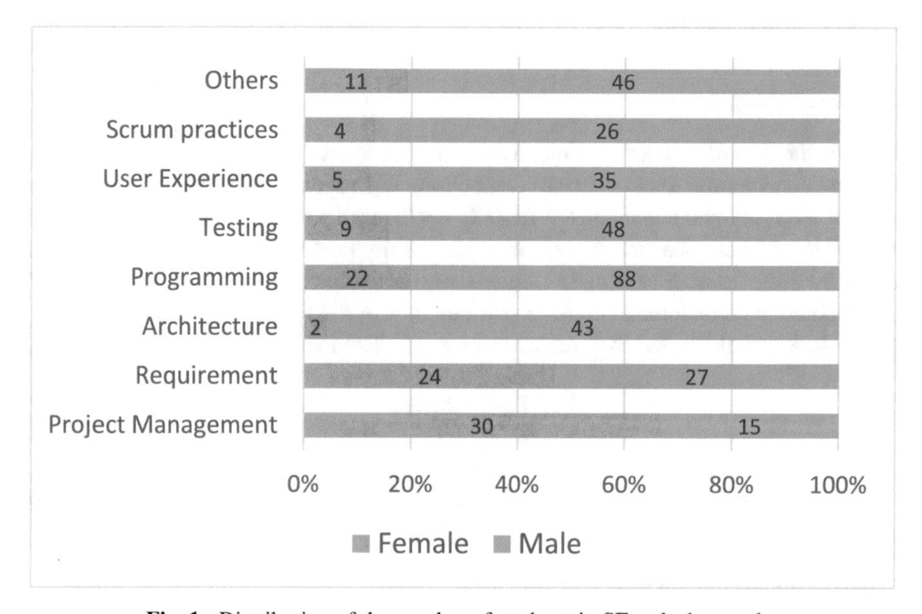

Fig. 1. Distribution of the number of students in SE tasks by gender

Project management- and requirement-related tasks were the most common tasks led by female students, which were performed by 70% and 56% of the female students, respectively. Whether assigned or voluntarily, female students expended considerable effort in coordinating team members and planning and tracking the execution of tasks. Female students were often involved in meetings with customers, participated in note-taking, and created user stories. The common activities performed by female members are illustrated in the following quote:

> "I try to have an overview and do all the management stuff, sending agenda, having an overview, especially in the first week I try to set the way we work, talking to customers. ... Mostly we made decisions together, everyone assigns themselves to what they want to do ... I worked a bit in the front-end development but I try to influence code convention... We have a very flat structure ... I don't think I have made any decision on my own."

A relatively large number of female students were actively involved in implementation (51% of female students). A female student typically took charge of a module (frontend/ backend), which was later integrated into a larger system. Here, a female student reports having a positive experience using control version systems to independently work on her task:

> *"... I just have to make the code to move the camera ... in Git we can work in different branches for the whole sprint in 2 weeks and then merge and still have it go smoothly..."*
>
> *"...guys who write the main code do testing as well..."*

Even though every team member was aware of ongoing tasks, what others were doing, and participated in all the tasks, there were only some members of a team that specially focused on a particular task:

> *"... front-end part is my area, but not as much as everybody else..."*
>
> *"I feel like I am a part of that but not in the real implementation team..."*
>
> *"Everyone participated in some way in implementation ... but when we have architectural problems there are two guys specialized in that ..."*

A total of 12% and 9% of the female students actively led user experience (design front-end layout, photos, etc.) and managed the Scrum methodology (i.e., Scrum master, product owner), respectively. Unit testing was often performed by the person who developed the code. The managing integration test, writing test reports, test cases, etc. were typically done by male students. Only 7% of the female students actively participated as test managers:

> *"for the last user testing we wrote the questionnaire ... I have done a lot of testing because I am an android tester, I am the main phone tester..."*
>
> *"...we have one guy that focuses on integration testing ..."*

Architecture is the area in which the female students were least active. This includes the technical plan for the whole system and certain specialized technologies (i.e., algorithms, artificial intelligence, system architecture, and security). We only had two cases in which female students actively led architecture activities:

> *"We used the Django framework ... it is more model-control-view model... I sort of decided which architecture we needed ..."*

In contrast, the distribution of male effort was more equal among the SE tasks. Male students often led programming (41%), testing (22%), and architecture activities (20%). A total of 16%, 13%, and 12% of males participated in user experience, requirements, and Scrum methodology, respectively. Project management was only performed by 7% of the male students.

Table 2 shows the probabilities that an SE task would be led by a female team in general, and in 2015, 2016, and 2017 specifically. The probability is benchmarked by the probability of having to pick a student and that student is female. In Table 2, we highlighted the tasks with a high chance of being led by a female. As can be seen in the table, the statistics are quite stable over the three years.

Project management was an activity area that was most likely to be led by females than males. Requirement and programming were also areas in which female students were more active, considering the percentage of female students in the class. Testing, user experience, Scrum practice, and architecture are knowledge areas in which females were underrepresented.

Table 2. Likelihood that a task is led by female students

Task	Overall	2015	2016	2017
Project management	67%	80%	64%	62%
Requirement	43.8%	44%	46%	42%
Programming	20%	19%	18%	22%
Testing	15.8%	15%	12%	20%
User experience	12.5%	9%	15%	13%
Scrum practice	13.3%	11%	11%	17%
Architecture	4%	0%	0%	14%

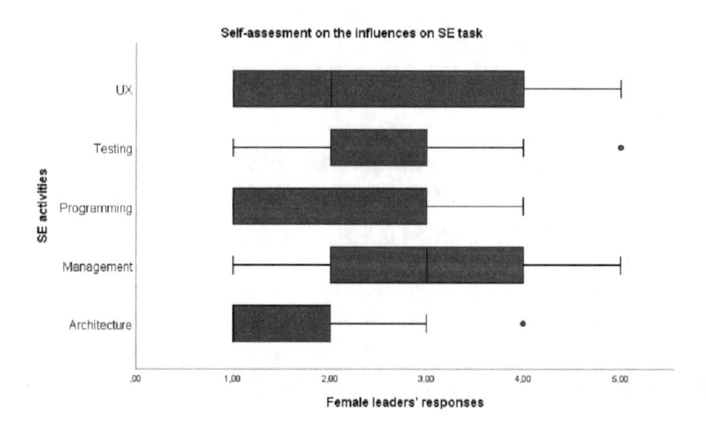

Fig. 2. Self-assessment on influences on SE activities

Figure 2 describes the boxplots of the responses from 19 female students regarding how they perceived their influence on different SE areas. The female students felt that they had a relatively better influence on management and user experience than on testing, programming, and architecture. Similarly, they felt that they had a greater influence on the areas of testing, programming, and management and relatively less influence on the areas of user experience and architecture. All median values were equal to or less than three (mid-value of a 5-point Likert scale). This might show that the female students did not generally think they influenced SE activities.

Takeaways:

- ***Female students were heavily involved in administrative, project management, and requirement activities***
- ***Female students were relatively little involved in Scrum method and architecture activities***
- ***Female students perceived themselves to have a limited influence on SE activities***

6 Discussion

6.1 The Role of Females in Software Project Courses

The literature has shown that students experience a different amount of learning in project courses [2, 5]. When the team organization emerges by itself, the activity profiles of female students are typically different from those of male students. Female students might learn more about project management or requirement engineering while male students might learn more about specific technical areas (e.g., architecture, testing, user experience, or Scrum practices). Such different learning experiences might come from a setting in which the female students are more associated with project managerial roles than with technical roles [5]. It is worth noting that the roles emerges and are self-assigned with students' teams. Whether assigned or self-selected, activity profiles represent how female students position themselves in a male-dominant software project.

An interesting observation is that the female students often participated in ordinary programming tasks, for example, the implementation of specific modules, front-end development (JavaScript, HTML, etc.), pair programming, unit testing, and code integration. Aligned with a previous finding [12], we found that female students were less active in complex technical problems (i.e., algorithms, artificial intelligence, system architecture, and security). The literature reports that there is a risk that female engineers will be allocated stereotypically female tasks [7, 9]. In our course, we found that past experience and knowledge were more critical factors for task assignments. However, gender did not seem to have a strong impact on team dynamics and team performance, contrary to the findings of previous research.

Although most of the student groups had a flat structure, many female students perceived that they had limited influence on activities, such as architecture or implementation. This might not be a particular issue for female students, as it can because of project management, how tasks are organized and implemented, how committed other team members are, etc. The difference in perception between female and male students here might relate to the different personality and individual characteristics, which has been observed in the literature [9, 11, 13].

6.2 Threats to Validity

We analyzed the threats to the validity of our study using the four categories of validity discussed by Runeson and Höst [20]: construct validity, internal validity, external validity, and reliability. We minimized construct validity by data triangulation. The student participation construct was captured by both quantitative data (survey) and qualitative data (interview, report), both subjective data (survey) and objective data (observation). Regarding external validity, we recommend the generalization of our observations to similar course settings (i.e., European university-level project courses in the fourth study year). Regarding reliability, our surveys are reused from previous studies, and they were compulsory for the students. We delivered the surveys and waited for the students to complete them. They could ask for clarification if any questions were not clear to them. Quantitative data were processed and analyzed in SPSS. The study protocol and data collection instruments are publicly available, so other researchers can repeat this research.

7 Conclusions

Complex features of software engineering project courses, including real clients, large teams, and flexible development paradigms, are hard to teach, but they have the potential to improve students' skills in many ways. Various course settings and their connections to learning outcomes have attracted research attention in SE education. However, female participation and learning is an area that has received relatively little attention. This study explores how female students participate in their teams in terms of engineering activities and team dynamics. We adopted a mixed research approach to collect three years of data from the Norwegian University of Science and Technology. We characterized the activity profiles of female students in a software project course. The areas of active involvement of female students are project management, requirement engineering, and administration. The areas that lack active involvement are architecture and Scrum methodology. Overall the female students have limited influences on SE activities.

The following step in this research would be to investigate other dimensions of team dynamics and their influence on team performance. The study could be replicated in other settings to improve the generalizability of the results. Instead of gender, future work might look at different biological characteristics, such as personality traits or individual motivation, in determining project performance.

Acknowledgement. This work has been partially supported by the EUGAIN COST Action CA19122 - European Network for Gender Balance in Informatics and by the NFR 295920 IDUN project.

References

1. Lingard, R., Barkataki, S.: Teaching teamwork in engineering and computer science. In: 2011 Frontiers in Education Conference (FIE), pp. F1C-1–F1C-5 (2011)
2. Bastarrica, M.C., Perovich, D., Samary, M.M.: What can students get from a software engineering capstone course?. In: Proceedings of the 39th International Conference on Software Engineering, Piscataway, NJ, USA, pp. 137–145 (2017)
3. Paasivaara, M., Vodă, D., Heikkilä, V.T., Vanhanen, J., Lassenius, C.: How does participating in a capstone project with industrial customers affect student attitudes?. In: Proceedings of the 40th International Conference on Software Engineering: Software Engineering Education and Training, New York, NY, USA, pp. 49–57 (2018)

4. Cleland-Huang, J., Rahimi, M.: A case study: Injecting safety-critical thinking into graduate software engineering projects. In: 2017 IEEE/ACM 39th International Conference on Software Engineering: Software Engineering Education and Training Track (ICSE-SEET), pp. 67–76 (2017)
5. Vanhanen, J., Lehtinen, T.O.A., Lassenius, C.: Software engineering problems and their relationship to perceived learning and customer satisfaction on a software capstone project. J. Syst. Softw. **137**, 50–66 (2018)
6. Marques, M., Ochoa, S.F., Bastarrica, M.C., Gutierrez, F.J.: Enhancing the student learning experience in software engineering project courses. IEEE Trans. Educ. **61**(1), 63–73 (2018)
7. Hirshfield, L., Koretsky, M.D.: Gender and participation in an engineering problem-based learning environment. Interdisciplinary J. Problem-based Learn. **12**(1) (2017)
8. Wolfe, J., Powell, B.A., Schlisserman, S., Kirshon, A.: Teamwork in engineering undergraduate classes: What problems do students experience?. In: Annual meeting of the American Society for Engineering Education, New Orleans, LA (2016)
9. Appelbaum, S.H., Audet, L., Miller, J.C.: Gender and leadership? Leadership and gender? A journey through the landscape of theories. Leadersh. Org. Dev. J. **24**(1), 43–51 (2003)
10. Grossman Philip, J., Komai, M., Jensen James, E.: Leadership and gender in groups: An experiment. Canadian J. Econ. Revue canadienne d'économique 48(1), 368–388, (2015)
11. Gramß, D., Frank, T., Rehberger, S., Vogel-Heuser, B.: Female characteristics and requirements in software engineering in mechanical engineering. In: International Conference on Interactive Collaborative Learning, pp. 272–279 (2014)
12. Dugan, R.F.: A survey of computer science capstone course literature. Comput. Sci. Educ. **21**(3), 201–267 (2011)
13. Fisher, M., Cox, A., Zhao, L.: Using sex differences to link spatial cognition and program comprehension software maintenance. In: 22nd IEEE International Conference on Software Maintenance, pp. 289–298 (2006)
14. Beckwith, L., et al.: Tinkering and gender in end-user programmers' debugging. In: Conference on Human Factors in Computing Systems, pp. 231–240 (2006)
15. Burnett, M., et al.: Gender differences and programming environments: Across programming populations. In: 2010 ACM-IEEE International Symposium on Empirical Software Engineering and Measurement (2010)
16. Berenson, S.B., Slaten, K.M., Williams, L., Ho, C.W.: Voices of women in a software engineering course: Reflections on collaboration. In: ACM J. Educ. Res. Comput. **4** (2004)
17. TDT4290 Compendium https://sbs.idi.ntnu.no/tdt4290
18. Society, I.C., Bourque, P., Fairley, R.E.: Guide to the Software Engineering Body of Knowledge (SWEBOK(R)): Version 3.0, 3rd ed. IEEE Computer Society Press, Los Alamitos, CA (2014)
19. Somech, A.: The effects of leadership style and team process on performance and innovation in functionally heterogeneous teams. J. Manag. **32**(1), 132–157 (2006)
20. Runeson, P., Höst, M.: Guidelines for conducting and reporting case study research in software engineering. Empir. Softw. Eng. **14**(2), 131 (2008)
21. Andersen, R., Conradi, R., Krogstie, J., Sindre, G., Sølvberg, A.: Project courses at the NTH: 20 years of experience. In: Díaz-Herrera, J.L. (ed.) CSEE 1994. LNCS, vol. 750, pp. 177–188. Springer, Heidelberg (1994). https://doi.org/10.1007/BFb0017613
22. Jaccheri, L., Pereira, C., Fast, S.: Gender issues in computer science: lessons learnt and reflections for the future. In: 2020 22nd International Symposium on Symbolic and Numeric Algorithms for Scientific Computing (SYNASC), pp. 9–16 (Sep 2020). doi: https://doi.org/10.1109/SYNASC51798.2020.00014
23. EUGAIN • COST ACTION CA19122 – European Network For Gender Balance in Informatics. https://eugain.eu/ (Accessed 18 Jan 2022)

Sustainability in Business: Change, Growth and Future Impact

Digital Technologies and Sustainability Paradoxes – An Empirical Study of a Norwegian Media Group

Anh Nguyen Duc(✉) and Birgit Leick

Department of Business and IT, University of South Eastern Norway, Bø i Telemark, Norway
{angu,birgit.leick}@usn.no

Abstract. Large companies are increasingly faced with the dilemma of balancing between their economic performance and sustainable environmental and social impact. The recent development of information and communication technology (ICT) has certain implications on achieving cooperates sustainability goals. Although both the theory and literature on sustainable development is rich, there is a gap on empirical cases which provide practical insights on the challenges for companies, particularly the technology-focused ones. Our objective is to explore the sustainability challenges for a large software technology company that operates platforms and digital services. We conducted an exploratory case study based on various sources of information. By matching the company's sustainability goals with the observed initiatives and product characteristics, we revealed insights on five different challenges at the organizational level. The study has implications for both research and practice on sustainability dilemmata in the ICT sector.

Keywords: cooperate sustainability · sustainability paradox · case study · ICT

1 Introduction

Sustainability has been one of the most frequently heard buzzwords when talking about future development, regardless of whether we are talking about the future of our society, businesses, nature or the planet itself. In the last ten years, organizations have shown their commitment on Corporate Social Responsibility (CSR) [1], Sustainable Development Goals (SDGs) [2] and sustainable development. Sustainability is defined as a development that takes into consideration both current and future needs, based on three so-called pillars: economic, environmental and social needs [3–5, 13]. Sustainable development has been a global topic [13] and it is widely considered as an important issue in high-tech sectors. In the era of digital transformation, building sustainability requires integrating technology and data from the very beginning [23]. It is a recent call for "*technology ecoadvantage*", a mindset about utilizing advanced technologies and digitized operations to develop sustainable business solutions [23, 24].

With the rising popularity of digital platforms, social media, the Internet of Things, big data, and artificial intelligence, ones would expect advancements in productivity,

V. Gupta et al. (Eds.): SSEBIM 2022, LNISO 62, pp. 53–62, 2023.
https://doi.org/10.1007/978-3-031-32436-9_5

connectivity, and infrastructures in industries and societies. However, it is also evidenced that technological innovations might have a range of dysfunctional impacts that are threatening social and political stability [6]. Controversial debates on how big tech companies access and track users' interests and preferences monetize through personalized advertising, influence people through controlling social media, and, in general, virtual technologies have impacted our way of thinking and sensing [6, 8].

A recent report shows only a few cases of tech companies that incorporated sustainability into their strategy and do not only talk about being more sustainable but act upon it [7]. Furthermore, empirical studies on the role of ICT in such companies is scarce. A few studies revealed *dilemmata* in adopting digital technologies for sustainable development [6–8]. In this paper, we explored a case of a large Norwegian company that offers more than 55 digital products and services with more than a billion monthly online visits. Its core business is stated as "*online marketplaces with a strong focus on sustainability*"[1]. We investigated how the company engages in sustainable development through its concrete strategies and evolutions of its digital brands (55 + brands). Through a theoretical lens of complexity and paradox, we revealed different challenges in realizing the company's sustainability goals. Our contribution is two-fold: (1) we illustrate the theories of paradox and complexity of corporate sustainability and (2), we borrow a conceptual framework to explain challenges in the sustainable adoption and operation of ICT in an online space.

The paper is organized as follows: Sect. 2 presents the contemporary literature on corporate sustainability from complexity and paradoxical perspectives. Section 3 provides our research approach and case description. Section 4 presents the result of our analysis. Section 5 discusses and concludes the paper.

2 Background

The last decade has shown several work documenting tensions between different sustainability objectives, which are interdependent and conflicting in the context of large companies [14–19]. Smith and Lewis identified four types of paradoxes: paradoxes of belonging, learning, organizing, and performing [15]. The *learning paradox* is characterized by the tensions in the shaping of new and destroying of old systems. The *belonging paradox* is characterized by the tensions of the individual against the collective where employees are likely to face opposing yet co-existing roles. The *organizational paradox* is characterized by the tensions of routine *vis-à-vis* change and of collaboration, *vis-à-vis* competition. The *performing paradox* is characterized by the tension of differing and conflicting demands of various stakeholders.

Hahn et al. pointed out that empirical research on corporate sustainability is still scarce [15], and revealed some case study research. Slawinski and Bansal examine how organizations attend to the tensions between short-term and long-term orientations around corporate sustainability [16]. Jay describes how decision-makers in a sustainable hybrid organization navigate paradoxical tensions between different organizational outcomes [17]. Ghadiri et al. investigate how Corporate Social Responsibility (CSR)

[1] https://www.schibsted.com/2022/02/02/facts-about-schibsted.

consultants manage the tensions between profit and social responsibility and find evidence for "paradoxical identity mitigation" where employees simultaneously embrace and distance themselves from competing demands [18]. Berger et al. identify companies that follow a syncretic stewardship model that tries to cater together economic, social, and environmental needs through constant balancing and negotiation [19].

Jaaron et al. conducted a case study about the adoption of Industry 4.0 in mitigating the intertemporal tensions of balancing the short-term and long-term sustainability objectives [20]. Gaim et al. described how the illusion of paradox embrace could trigger dysfunctional behaviors in the case of Volkswagen [21]. Acquier et al. showed how the platform economy has promises and paradoxes.

Notwithstanding this body of literature, there are a few cases where the sustainability dilemma is explicitly discussed in the connection to digital technologies and linked to specific company cases.

3 Research Methodology

The study starts from a theoretical viewpoint on the definitions and realization of sustainable goals in large companies. In a context of a course in Master of Sustainability[2], we looked for empirical evidence on sustainable development. Certain criteria were considered for the search, (1) the case should be a clear example of "sustainability in action", (2) the case should be highly relevant to ICT adoption as a requirement of the course, and (3) the case should be educational. Several options were identified through the authors' professional network, and we decided to select the case with sufficient details to analyze and report in this work.

3.1 The Case

Schibsted ASA is one of the largest Nordic media groups that operates in both Norwegian and international markets. The cooperates' s headquarter is in Oslo, Norway. The group had a turnover of MNOK 14,623 in 2021 [11]. Schibsted's operations are divided into three business areas: marketplaces, news media and financial services and ventures[3]. The group owns several platforms, such as FINN, Blocket, Tori, Oikotie and DBA that connect millions of buyers and sellers every month. FINN[4] is the flagship of Schibsted and the leading website for online marketplaces in Norway. Here, they are market leaders in the vehicle, second-hand goods, real estate and job categories. In 2016, FINN had an operating income of NOK 1,650 million. The website is one of the most popular in Norway measured by internet traffic. Moreover, Schibsted owns the leading newspapers Aftenposten and VG in Norway and Aftonbladet and Svenska Dagbladet in Sweden. All of them are popular online newspapers with subscription models. Moreover, the

[2] https://www.usn.no/english/academics/find-programmes/master-of-sustainability-management-exchange/sustainability-management-master-S-level-1. The commitment of the master students in this class who had partly worked on this company case in their course assignments and examination is gratefully acknowledged.

[3] https://www.schibsted.com/about/.

[4] https://www.finn.no/.

group invests in digital companies that empower consumers with the aim of businesses internationally and growth to become market leaders. Examples of such investments are Blocket, Prisjakt, TV.nu and Lendo.

In term of its business model, the group's operating income stems from three major sources: (1) payments of customers, readers and subscribers for paper newspapers and online newspapers, (2) commercial advertisements, which are typically pay-per-click for advertisers, on websites and in newspapers, and (3) online classified advertisements where the income mainly comes from professional customers such as estate agents and employers.

Schibsted considers itself in a position to make an important societal and environmental impact. Written from their website, second-hand trade is an important contribution to a sustainable world. And it is claimed that Schibsted's marketplaces saved 20.5 million tons of greenhouse gas [12]. The company's strategy relies on the leverage of digitalization to create sustainable solutions for many social issues. The company has defined for itself sustainability aspects, long-term performance ambitions and short-term targets. For instance, in 2015, Schibsted initiated a program called "the Second-Hand Effect", making a visible calculation to show the environmental benefits form second-hand trade. Schibsted wants to raise awareness about the environmental benefits of reusing and repairing goods and minimizing waste.

3.2 Data Collection and Analysis

The case was part of a Master's course "Process, Product and People" in the autumn term of 2021 in the School of Business of the University of South-Eastern Norway. In the course context, sustainable development in Schibsted was investigated and analyzed. We invited the head of sustainability at Schibsted to give a keynote lecture for our master students. The lecture was followed up by a discussion section with the students. By including this material in the analysis, this paper is informed by insights into how sustainability is defined and realized through concrete initiatives in different units of the case company. We understood their strategies and adopted technologies and business development toward sustainability. We also included in the analysis the annual report 2021 [12] and the sustainability report 2021 [11] from the company. Including these reports in this analysis provides further insights into the values and actions regarding diversity that Schibsted wants to promote externally of the company. We also included in our analysis some of the media coverage of Schibsted in the context of digitalization and sustainability in a selection from 2015 to today. These are relevant external factors that also play a role in Schibsted's motivation to transform its organizational development. Our analysis is done based on all materials of the case that is accessible to us. Other sources of data, such as observations or interviews with more stakeholders, might provide even better insights into the case. However, it is also noted that studying a large organization requires a thorough understanding of strategic documents and company reports, which often appear in public format. Besides, our findings illustrated and strengthened the paradoxical view on the implementation of sustainability goals in cooperates. A study that occurred in different contexts might offer different views from this work.

4 Findings from the Case Company

By analyzing the keynote lecture and associated materials from complexity and paradoxical aspects, we reveal four challenges towards implementing Schibsted's sustainability goals in their actions: (1) Gaps between their strategical sustainability targets and their operationalization (Sect. 4.1), (2) Communication of sustainability at the organizational level (Sect. 4.2), (3) Tensions in achieving concurrent sustainable goals and (4.3) Complexities of driving sustainability at an organizational level (Sect. 4.4) and Paradox perspectives in Schibsted's sustainability approach (Sect. 4.5).

4.1 Gaps Between Their Strategical Sustainability Targets and Their Operationalization

In defining the company's sustainability scope, three aspects were highlighted as unique elements to Schibsted: (1) empower circular and sustainable consumption, (2) independent and high-quality journalism and (3) empower people to make informed choices. We synthesized the analysis of company's sustainability report 2021, media covers and our analysis of their marketplace product – FINN.

FINN has a great potential for contributing to empowering the circular and sustainable consumption. Through ICT, listings, communication, and payment is executed through FINN's own platform (Fig. 1a). This platform, therefore, provides a solution for circular consumption, leading to more environmentally friendly consumption. Schibsted has even made a calculation that shows how much potential the circular consumption in all of their digital marketplaces have combined in a so-called "the Second-Hand Effect" strategy. However, it can be argued that Schibsted's goals with "The Second-Hand Effect" are not actually sustainable goals. Why? "The Second-Hand Effect" that Schibsted is marketing as their greening of the consumer value only sets out to measure the quantity of second-hand consumption, which means little unless the overall production and consumption of new products also decreases. When it comes to finding the right ways to measure sustainability, it is important to think through what the systematic and strategic goals put forth will produce in terms of results and evaluate to what extent those results can be considered sustainable.

4.2 Communication of Sustainability at the Organizational Level

Since Schibsted has a lot of digital assets at its disposal, the storytelling of the company's sustainability efforts can be made even more compelling with modern tools of data visualization. The hard efforts they have put in the background can be made more visible and transparent for people who try to understand and align with their mission through the use of innovative data visualization methods. An array of 53 strong brands and many countless men and women leading those brands are useful resources. They can be put to best service for sustainability by fostering intra-company partnerships within the group itself because partnerships need not be sought outside the family. They can be groomed inhouse.

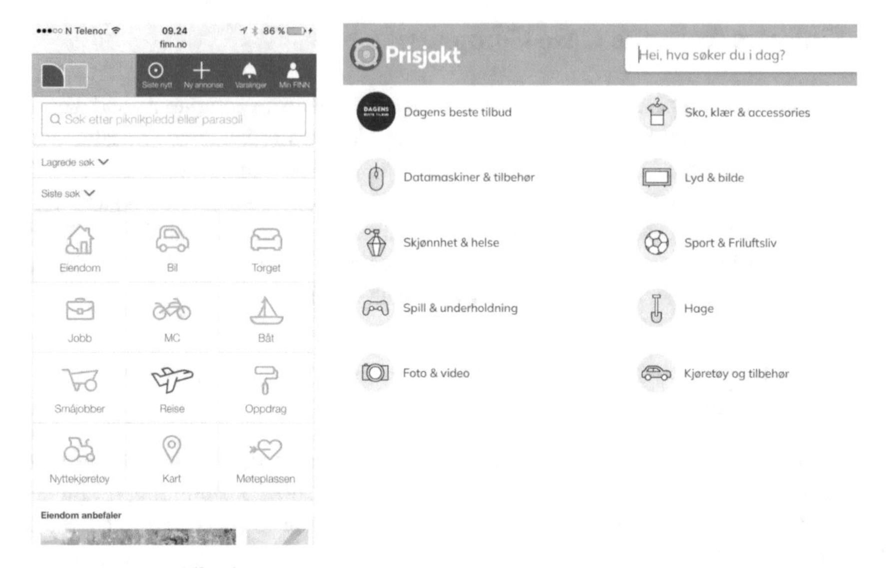

Fig. 1. a. Finn.no interface b. Prisjakt website interface

4.3 Tensions in Achieving Concurrent Sustainable Goals

Schibsted can benefit from using blockchain technology for circular and responsible consumption. Marketplaces and classifieds could then make a system of keeping track of the number of times for each unique item being distributed, which enables Schibsted to calculate their actual second-hand effect more precisely. However, there is a known drawback with blockchain towards sustainability: the enormous use of energy for computing (Morabito, 2016). If you try to reduce the sustainability impact in areas such as responsible consumption, you might end up increasing your impact in another sustainability areas or with regard to other sustainability goals, here exemplified by energy consumption. Schibsted needs to weigh the envisioned positive benefits of their sustainability strategies with the potential negative sustainability impact that their actions could produce with every movement and strategy they take.

Another highlighted strategy from Schibsted is to empower people to make informed decisions. Schibsted is using advanced cloud technology, big data and machine learning in order to provide consumers with the service called prisjakt.no[5] (Fig. 1b) where the customer can make an informed decision on the lowest price available for his or her purchase. It might be contradictory to the company's future-fit way of thinking to support consumers on finding a best bargain. If they want their ICT products and services to green the consumer, Schibsted would have to change prisjakt.no to also be informing the consumer of products' full life cycle from cradle to grave.

4.4 Complexities of Driving Sustainability at the Organizational Level

Schibsted is a huge corporation that is packed with complexities that come with different types of problems and issues. Externally, the competing marketplace is full of complex

[5] https://www.prisjakt.no/.

aspects imposed by other large corporations, which include financial performance, profitability, customer satisfaction, sales and services, product placement and visibility of branding. Schibsted is oriented to present itself as a sustainable company and build a new strategy for branding in order to become more eco-friendly and competitive in the marketplace. Internally, Schibsted is operating with a large network of 55 digital brands. Sometimes complexity could arise when different companies apparently are similar but have different ideas about tasks or projects. In every company, the policies and rules are aligned with the company's goals, and also in most cases, the companies are located in various countries with different working cultures and organizational backgrounds. All of those reasons can make company structure and process complex and unstable.

Sustainability in Schibsted complexities could furthermore come out when it comes to projects which are less oriented on profits but oriented more on the environment and social performance. Such cases can generate a lot of discussion and arguments inside the company between the management/financial team and shareholders. In 2019 and 2020, when Schibsted started to put most of the effort in sustainability, the management team got an agreement from shareholders to build up a better sustainable environment for the company, which led to better financial performance, which in earlier years was impossible.

4.5 Paradox Perspectives in Schibsted's Sustainability Approach

It is written in the Schibsted's sustainability report that "*Sustainability can no longer be written off as a mere lifestyle choice among consumers, or a legal commitment among governments. Environmental and social sustainability are increasingly integral parts of a business' financial sustainability. Sustainability earns you the right to play. For Schibsted, sustainability is at the very core of our business model.*" The view expressed in this statement emphasizes the strategy of environmental and social aspects as '*a part of financial sustainability*' which is contrary to the paradox perspective as it requires all three goals to have equal footing as presented earlier. We cannot find in our analyzed materials anywhere a presentation of solutions for this probable tension between the goals. We discussed the four types of paradoxes [14] relating to Schibsted' sustainability goals:

Organizing Paradox: Each sub-units within Schibsted has individual financial targets whereas as an organization they have a collective ambition for sustainability. This can be seen as an organizing paradox for Schibsted's sustainability management. The company is trying to manage the resulting tension through group-level policies that all subsidiary businesses must adhere to.

Performing Paradox: General managers of various companies are responsible for the implementation of sustainability plans in respective companies. Similarly, Schibsted also produces "*Sustainability Change-makers*" [10] from within the employees. They are also obliged to meet traditional performance metrics alongside sustainability indicators. This puts the managers in a paradoxical position where they are divided into delivering results in opposite goals.

Belonging Paradox: Acquisition and development of start-up companies is a key growth strategy for Schibsted. They are always seeking for "*disruptive, scalable and innovative business models that create unique value*" and over years have successfully acquired several brands like Lendo, Let's Deal, *etc.* [9] They will also face an eventual exit from certain businesses. A belonging paradox arises among the managers of newly acquired companies that have to transform their values to align with Schibsted's sustainability-oriented values. The paradox also arises among the managers of newly exited companies that have to transform their Schibsted aligned values to ones that enable the newly exited companies to survive in the market.

Learning Paradox: Schibsted operates in a future-oriented business environment with high levels of uncertainty such as in the field of data journalism and venture investment. This means the skills and capabilities developed over time become redundant in a very duration. This creates a perpetual learning paradox where the organization needs to learn rapidly but does not know what and how to learn.

5 Discussions and Conclusions

We observe a clear opportunity towards dealing with some sustainable goals at a local level. A tech company can leverage advanced technologies, *i.e.*, data visualization, Internet of Things, and Big data for tracking products being manufactured, shipped, and delivered, and present to its consumers a broader perspective of the impacts each purchase or consumption has. Making sustainability impacts "real" by making them visual rather than theoretical could provide a push towards more sustainable consumption. With the need for the recommendations given through the marketplace *Prisjakt* to become more conscious to increase its ability to green the consumer, the need also increases for AI technology to develop further to better implement the complexities in sustainable decision-making.

To fully operate this benefit of technology, companies like Schibsted should have a clear view of their potential challenges in balancing all sustainability's pillars. An organization can adopt some strategies from a paradox perspective to manage the integration of sustainability throughout the organization. One could start with the identification of paradoxical tensions related to organizing, performing, belonging and learning that are currently manifest in various aspects of its operation. Many such tensions could be latent at present but can surface in future when the environmental context changes. It could be a good idea to connect different sectors and companies together and take perspectives of long-time horizons to identify management strategies to accept and live with paradoxes or partially resolve some components of paradoxes with iterative and dynamic strategies. Accepting that managing sustainability in a modern large enterprise is complex and paradoxical with persistent tradeoffs between various significant elements of business will help explore and realize even more areas for sustainability gains.

This study presented a case study about challenges in the implementation of sustainability goals in Schibsted, and revealed that the actual sustainability journey might be different from the ones portrayed in the public reports. The achievement of SDGs can be supported by the strategical adoption of ICT. The need for data related to social, economic and environmental goals is immense as the major policies developed by government and business are mostly framed according to the SDGs. It might be needed to have revisit on the goals, metrics, and action plan and map those with the actual outcomes.

References

1. McWilliams, A., Siegel, D.: Corporate social responsibility: a theory of the firm perspective. Acad. Manag. Rev. **26**(1), 117–127 (2001)
2. https://sdgs.un.org/goals
3. Bansal, P., Des Jardine, M.: Business sustainability: it is about time. Strateg. Organ. **12**(1), 70–78 (2014). https://doi.org/10.1177/1476127013520265
4. Ehnert, I., Parsa, S., Roper, I., Wagner, M., Muller-Camen, M.: Reporting on sustainability and HRM: a comparative study of sustainability reporting practices by the world's largest companies. Int. J. Human Resour. Manage. **27**(1), 88–108 (2016). https://doi.org/10.1080/09585192.2015.1024157
5. A.B. Carroll Business and Society Little Brown & Company, Boston, MA (1981)
6. Arogyaswamy, B.: Big tech and societal sustainability: an ethical framework. AI & Soc. **35**(4), 829–840 (2020). https://doi.org/10.1007/s00146-020-00956-6
7. Kiron, D., Unruh, G., Reeves, M., Kruschwitz, N., Rubel, H., Zum Felde, A.M.: Corporate Sustainability at a Crossroads. MIT Sloan Manage. Rev. **58**(4) (2017)
8. Sætra, H.S., Coeckelbergh, M., Danaher, J.: The ai ethicist's dilemma: fighting big tech by supporting big tech. AI and Ethics 1–13 (2021). https://doi.org/10.1007/s43681-021-00123-7
9. Schibsted sustainability report (2021). https://static.schibsted.com/wp-content/uploads/2022/04/07102011/Sustainability-Report-2021-FINAL.pdf
10. Schibsted sustainability strategy. https://schibsted.com/sustainability/our-sustainability-strategy. Access dated: Aug 2022
11. Schibsted annual report (2021). Available online at https://static.schibsted.com/wp-content/uploads/2022/04/07164038/Schibsted-Annual-Report-2021.pdf
12. Schibsted sustainability. Available online at https://schibsted.com/sustainability/. Access dated: Aug 2022
13. Brown, B.J., Hanson, M.E., Liverman, D.M., Merideth, R.W.: Global sustainability: toward definition. Environ. Manage. **11**, 713–719 (1987). https://doi.org/10.1007/BF01867238
14. Smith, W.K., Lewis, M.W.: Toward a theory of paradox: a dynamic equilibrium model of organizing. AMR. **36**, 381–403 (2011). https://doi.org/10.5465/amr.2009.0223
15. Hahn, T., Figge, F., Pinkse, J., Preuss, L.: A Paradox perspective on corporate sustainability: descriptive, instrumental, and normative aspects. J. Bus. Ethics **148**(2), 235–248 (2017). https://doi.org/10.1007/s10551-017-3587-2
16. Slawinski, N., Bansal, P.: Short on time: Intertemporal tensions in business sustainability. Organ. Sci. **26**(2), 531–549 (2015). https://doi.org/10.1287/orsc.2014.0960
17. Jay, J.: Navigating paradox as a mechanism of change and innovation in hybrid organizations. Acad. Manag. J. **56**(1), 137–159 (2013). https://doi.org/10.5465/amj.2010.0772
18. Ghadiri, D.P., Gond, J.-P., Bre‘s, L.: Identity work of corporate social responsibility consultants: Managing discursively the tensions between profit and social responsibility. Discourse Commun. **9**(6), 593–624 (2015)
19. Berger, I.E., Cunningham, P.H., Drumwright, M.E.: Mainstreaming corporate social responsibility: developing markets for virtue. Calif. Manage. Rev. **49**(4), 132–157 (2007)
20. Jaaron, A.A.M., Backhouse, C.J.: Can Industry 4.0 hold the answer for mitigating intertemporal tensions in sustainable manufacturing? A conceptual model. Technol. Anal. Strategic Manage. **0**, 1–13 (2021). https://doi.org/10.1080/09537325.2021.1989399

21. Gaim, M., Clegg, S., Cunha, M.P.: e: Managing impressions rather than emissions: volkswagen and the false mastery of paradox. Organ. Stud. **42**, 949–970 (2021). https://doi.org/10.1177/0170840619891199
22. Acquier, A., Daudigeos, T., Pinkse, J.: Promises and paradoxes of the sharing economy: an organizing framework. Technol. Forecast. Soc. Chang. **125**, 1 (2017). https://doi.org/10.1016/j.techfore.2017.07.006
23. Close, K., Faure, N., Hutchinson, R.: How Tech Offers a Faster Path to Sustainability; Boston Consulting Group: Boston. MA, USA (2021)
24. Hajishirzi, R., Costa, C.J., Aparicio, M.: Boosting Sustainability through digital transformation's domains and resilience. Sustainability. **14**, 1822 (2022). https://doi.org/10.3390/su14031822

The Leading Locations of Information Technology (IT) Jobs in South Africa

Oluwaseun Alexander Dada[1,2](✉), George Obaido[3], Ibomoiye Domor Mienye[4], and Kehinde Aruleba[5]

[1] Department of Computer Science, University of Helsinki, Pietari Kalmin katu 5, 00560 Helsinki, Finland
alexander.dada@helsinki.fi

[2] The School of Software, 18 Kessington Street, Lekki 101245, Lagos, Nigeria
alexander.dada@schoolofsoftware.net

[3] Department of Computer Science and Engineering, University of California, San Diego, USA
gobaido@eng.ucsd.edu

[4] Department of Electrical and Electronic Engineering Science, University of Johannesburg, Johannesburg, South Africa
ibomoiyem@uj.ac.za

[5] Department of Information Technology, Walter Sisulu University, Mthatha, South Africa

Abstract. One of the main factors influencing job satisfaction for employment that require onsite presence is location. We analysed 3,355 jobs in Information Technology (IT) published on LinkedIn's South Africa website with the aim of identifying the leading locations for IT jobs in South Africa, as well as the names and industry sectors of the recruiting companies. Three (of the nine) provinces accounted for 91% of all IT job vacancies in South Africa, namely Gauteng (46.7%), Western Cape (39.4%) and KwaZulu-Natal (4.8%). Analysis of job distribution by cities revealed that Cape Town was home to 33% of all IT jobs in South Africa, followed by Johannesburg (24%) and Pretoria (3%). The Northern Cape province had the fewest IT job openings, followed by the Free State and Limpopo. Overall, the three most in-demand IT jobs were DevOps Engineers, Software Engineers and Data Engineers. Among the companies with the largest number of IT jobs, DigiOutsource came first, followed by takealot.com and impact.com. Five out of the top ten IT jobs advertised came from companies in the Staffing & Recruiting industry. The conclusions have practical implications for a variety of stakeholders, including businesses, governments, educational institutions, IT students and IT professionals.

Keywords: IT jobs · South Africa · Job locations · Hiring companies · LinkedIn · Text mining · IT students · IT professionals

1 Introduction

The COVID-19 pandemic has caused a major shift in how people think about 'work' - particularly about where they work. Research has shown that job loca-

V. Gupta et al. (Eds.): SSEBIM 2022, LNISO 62, pp. 63–74, 2023.
https://doi.org/10.1007/978-3-031-32436-9_6

tion is one of the most important factors individuals consider when looking for a new job or evaluating a job offer [1]. About 60% of the respondents in a study conducted in the United Kingdom placed more value on job location than on financial remuneration (e.g., salary) [8]. According to the results of another study carried out in Canada, 68% of the survey participants selected job location as one of the primary factors they considered when making career decisions [9]. We analysed 3,355 jobs in Information Technology (IT) published on LinkedIn's South Africa website with the aim of identifying the leading locations for such jobs, as well as the recruitment companies. To achieve this objective, we set out to answer the following research questions. (1) Which IT jobs are in the highest demand in South Africa? (2) Which companies advertised the most job vacancies in IT in South Africa? (3) Which locations (provinces and cities) had the largest number of job vacancies in IT in South Africa?

The findings from this study will add to the existing literature on IT opportunities in South Africa. Given that location is an essential part of the decision-making process among individuals considering job offers, our findings will provide applicants with a good understanding of how job opportunities in IT vary across locations in South Africa.

The rest of this paper is structured as follows. Section 2 comprises the literature review. The methodology used in the investigation is explained in Sect. 3. Section 4 presents the results and the discussion, and the implications are discussed in Sect. 5. The industrial application of the results is dealt with in Sect. 6. Section 7 concludes the paper.

2 Literature Review

2.1 Why South Africa?

South Africa is arguably the most industrialised country in Africa and one of the largest emerging market economies in world [10]. The country is sometimes referred to as the gateway into Africa because of its well-developed financial markets and fast-paced rising economy. It has also evolved into a vibrant and entrepreneurial investment location with many unique competitive advantages. The stunning beauty, unusual wildlife, and diversified culture of South Africa have helped the country develop its reputation as one of the best tourist attractions in the world [10]. Furthermore, South Africa possesses Africa's strongest education system, and it is one of the few African countries with world-class universities. This encourages the growth of a contemporary economy that is diversified and inventive [11].

2.2 South African Job Opportunities for IT

South Africa's unemployment rate had reached 34.4% as of August 24, 2021, the highest among 82 countries analysed by Naido [12]. However, research conducted by CareerJunction [22] shows that there has been an increase in demand for

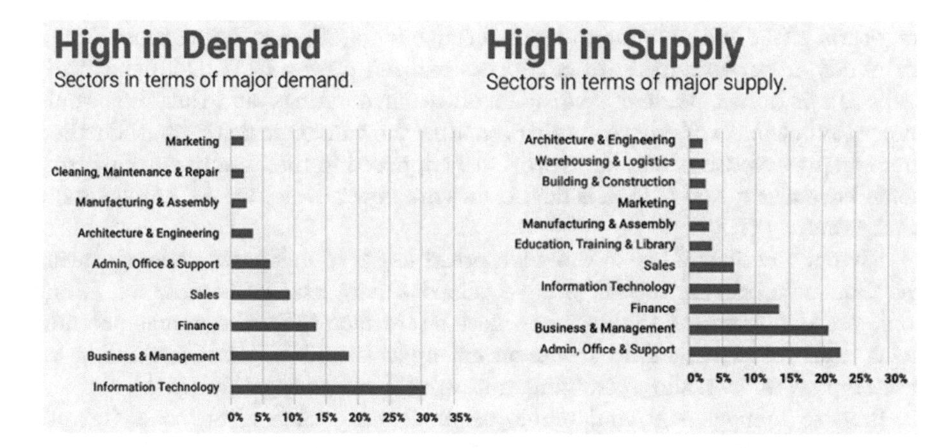

Fig. 1. The employment landscape in South Africa [22]

labour, specifically in the field of technology. CareerJunction is one of the biggest online recruitment platforms in South Africa, with millions of registered job applicants and thousands registered employers in its database.

Figure 1 gives a good indication of the demand and supply of labour in South Africa. The demand for IT workers significantly surpasses the supply, and it is obvious that there is a shortage of IT workers and potentially less competition for IT jobs.

As companies grapple with the shortage of skilled workers, it appears that job searchers are beginning to place a higher value on where they work than on monetary compensation. According to a study [22], 68% of survey respondents said location was the most important factor in deciding whether or not to accept a job offer. The findings of another study [1] revealed that employment location was the crucial factor influencing why people stayed in their current jobs. The scholars claimed that 57% of respondents valued their workplace more than their salary [1].

2.3 Post-COVID-19 and the New Normal

Since the outbreak, the pandemic has had a negative influence on practically every area of the global economy [3,5,6]. In response, many companies and employees have been compelled to adopt new ways of working that are likely to stick after the pandemic. The following are aspects of traditional work life that have been impacted by the pandemic and may remain even afterwards.

Remote work: The demand for remote work has skyrocketed on LinkedIn recruitment platform around the world since the beginning of the pandemic. According to BBC news, the number of job searches using the "Remote" filter on LinkedIn has climbed by 60% between March and November 2020. It was also reported that percentage of Remote Job Applications, during the same period, increased about 2.5 times internationally. This is expected to have a significant long-term effect on the labor and global economy [4,20]. According to an empirical study, a large majority of employees realized during COVID-19 that their tasks can be done outside of traditional office surroundings, and that they would prefer to continue working remotely even after the pandemic has passed. Furthermore, many executives who were previously opposed to their teams working from home have discovered that it is doable and are now advocating for employees to work remotely [7].

Virtual meetings: Due to the widespread usage of videoconferencing during the pandemic, virtual meetings have gained a new level of acceptance. Many commercial operations that used to necessitate face-to-face meetings are now being done in virtual mode. Common examples include digital onboarding for new employees, workshops/training, and workplace get-togethers [21].

Remote management and employee monitoring: According to a Gartner study, 16% of companies are increasingly employing technology to monitor their employees including virtual clocking in and out, tracking work computer usage, and monitoring employee emails or internal communications/chat. While some businesses track productivity, others track employee engagement and well-being in order to gain a better understanding of the employee experience [19].

3 Methodology

To achieve our goals, we extracted publicly available information on Information Technology (IT) jobs posted on LinkedIn in South Africa, using the keyword "IT" on the portal (https://za.linkedin.com).

3.1 Data Pre-processing

The dataset was extracted between May 2021 and August 2021 via web scrapping, using Python programming packages (Beautiful Soup, requests and pandas). The raw data contained 5,251 records. Following the removal of duplicates, 3,355 records were left, but these were in rows/columns. Some of the adverts were for multiple positions (IT and non-IT jobs), and we decided to remove them from the dataset to improve the analysis. The data represented the following fields: 'job title', 'company', 'location' and 'industry'.

Some of the records were further refined to make them more descriptive and easier to process. For instance, because we were interested in locations as provinces and cities, the original values were in the form: "city, province, country" (e.g., 'Kimberley, Northern Cape, South Africa'), "province, country" (e.g., 'Mpumalanga, South Africa') and 'country' (e.g., 'South Africa'). The names of the metropolitan areas were used instead of the name of the provinces in some records, which we changed accordingly: for instance, "Johannesburg Metropolitan Area" was changed to "Gauteng", "Port Elizabeth Metropolitan Area" to "Eastern Cape" and "Durban Metropolitan Area" to "Kwazulu-Natal". In addition, some locations had the value "Hide Job", in which case we simply replaced them with "South Africa". Two more fields (provinces and cities) were added at the end of the data pre-processing stage.

4 Results and Discussion

The analysis identified 2,362 unique job titles, 659 unique company names and 275 unique industries, as well as 118 unique locations that were further refined to 103 cities across the nine provinces of South Africa.

4.1 RQ 1: Which IT Jobs Are in the Highest Demand in South Africa?

As shown in Fig. 2, DevOps Engineer is the most common job title, followed by Software Engineer (in second place) and Data Engineer in the third position.

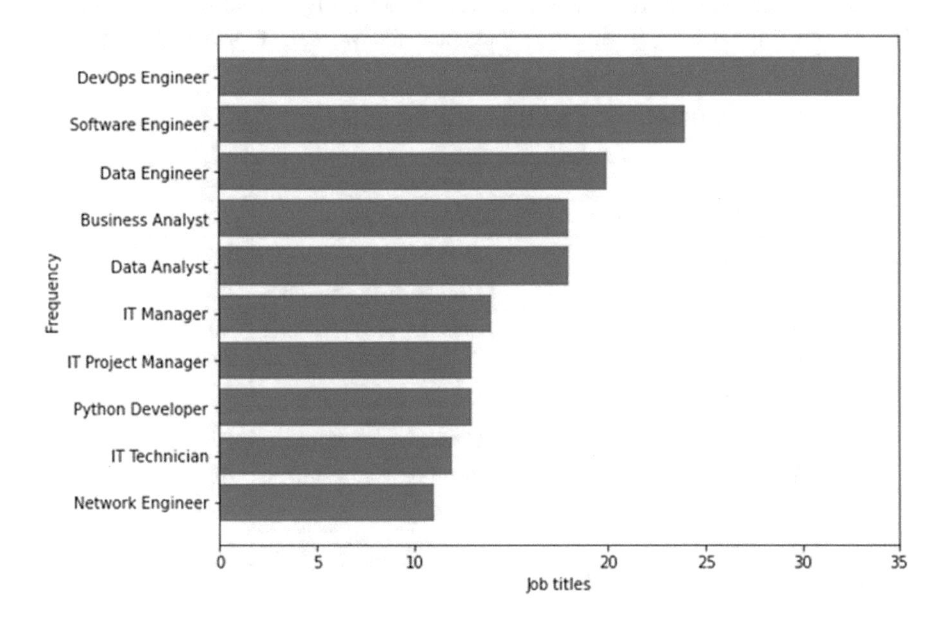

Fig. 2. The top ten job titles in terms of frequency

Of these 10 job titles, three are within IT support (DevOps Engineer, IT Technician and Network Engineer), two belong to the "data" function (Data Engineer and Data Analyst), two to management (IT Project Manager and IT Manager), two to software development (Software Engineer and Python Developer) and one to "business" (Business Analyst). Based on frequency, the job functions with the largest numbers of job vacancies are IT support, software development and "data".

Our finding that DevOps engineer was the leading IT job in South Africa is consistent with information posted on LinkedIn in 2018 indicating that, in terms of recruitment, DevOps engineer was, overall, the top job across all industries [13]. One important characteristic of DevOps engineer is the ability to combine technical skills (e.g., coding, problem solving) with soft skills (e.g., effective collaboration, presentation of ideas). The key responsibilities for a DevOps engineer include automation, configuration management, Docker container usage, Continuous Delivery/Continuous Integration (CD/CI). This will help companies to become more agile [2] and effectively to changes in their markets. In doing so, companies will increase their time-to-market and overall projects' performance in production [14].

4.2 RQ 2: Which Companies Advertised the Most Job Vacancies in IT in South Africa?

Table 1 shows that DigiOutsource (industry: IT & Services) was the company with the largest number of IT job vacancies, followed by takealot.com (industry: Retail) and impact.com (industry: Computer Software) was in third place.

Table 1. The top ten companies with the most IT job adverts.

Rank	Company	Industry	Frequency
1	DigiOutsource	IT & Services	112
2	takealot.com	Retail	108
3	impact.com	Computer Software	93
4	Goldman Tech	Staffing & Recruiting	90
5	CareerJunction	Staffing & Recruiting	89
6	Luno	Financial Services	84
7	Discovery Limited	Financial Services	76
8	Black Pen Recruitment	Staffing & Recruiting	62
9	Unique Personnel	Staffing & Recruiting	57
10	HR Genie	Staffing & Recruiting	56

Excluding the top three, five (5) of the remaining seven in the top ten (10) companies specialised in Staffing & Recruiting, and two (2) were from Financial Services.

DigiOutsource is an online and mobile e-commerce enterprise and operates as an IT out-sourcing firm making it possible for its employees to work remotely from any location around the world [15]. On closer inspection, half (50%) of the leading IT jobs in South Africa seem to be handled by recruitment companies. It is not clear why this is so, or if it is now the trend.

4.3 RQ 3: Which Locations (Provinces and Cities) had the Largest Number of Job Vacancies in IT in South Africa?

We looked at locations in terms of provinces and cities. As Fig. 3 shows, half of the jobs were located in only three of South Africa's nine provinces: Gauteng (1,568 jobs, frequency: 46.7%), Western Cape (1,321, frequency: 39.4%), and KwaZulu-Natal (162, frequency: 4.8%) as the distant third.

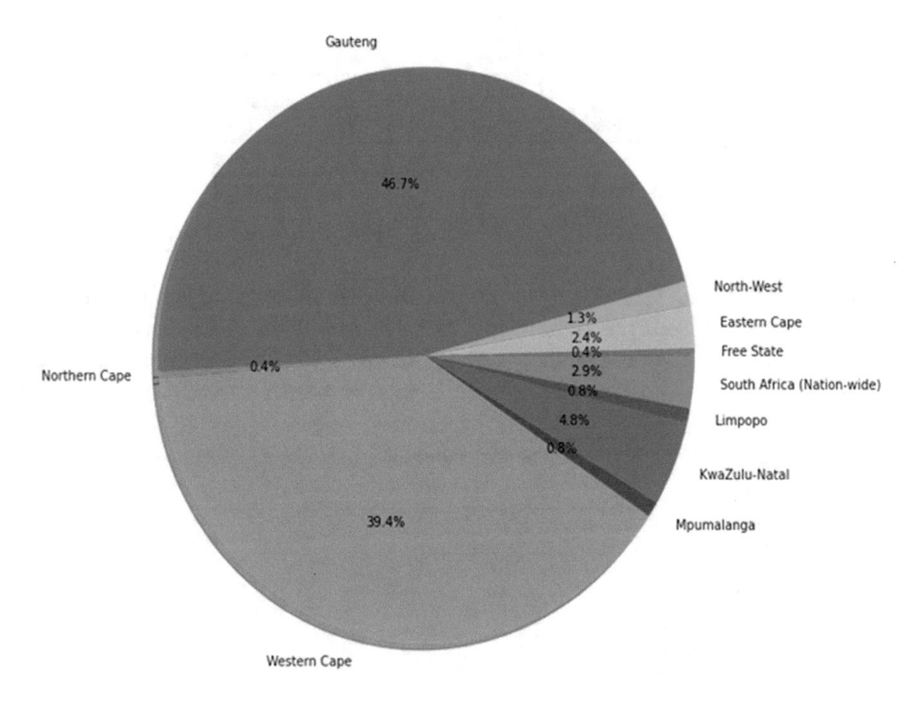

Fig. 3. Percentage distribution of IT jobs across South African (and national) provinces

The province with the least number of IT job vacancies was Northern Cape (12, frequency: 0.4%), followed by Free State (14, frequency: 0.42%) and then Limpopo (27, frequency: 0.8%). In terms of composition, Gauteng and the Western Cape accounted for over 86% of all IT jobs advertised in South Africa. Along with KwaZulu-Natal, the three provinces constitute about 91% of all published job vacancies in IT. It is not surprising that Gauteng province, which is the economic hub of South Africa, has the largest number of IT vacancies. It is home to top-500 companies and boasts many technology-intensive projects initiated by both governments and multinational corporations.

Table 2. Ranking, frequency and percentages of the top ten cities with the highest number of job vacancies in IT.

Rank	Cities	Province	Frequency	Percentage
1	Cape Town	Western Cape	1101	32.8%
2	Johannesburg	Gauteng	794	23.7%
3	Pretoria	Gauteng	107	3.2%
4	Durban	KwaZulu-Natal	90	2.7%
5	Midrand	Gauteng	88	2.6%
6	Sandton	Gauteng	80	2.4%
7	Centurion	Gauteng	69	2.1%
8	Stellenbosch	Western Cape	61	1.8%
9	Bellville	Western Cape	55	1.6%
10	East Rand	Gauteng	32	1.0%

On the city level, Table 2 shows the top 10 cities in terms of job vacancies in IT throughout the country, amounting to 2,477 vacancies (74% of the total IT jobs in the country). As Fig. 4 shows, Cape Town (1,101 jobs) topped the list by a significant margin, hosting 33% of all IT jobs in South Africa. Johannesburg

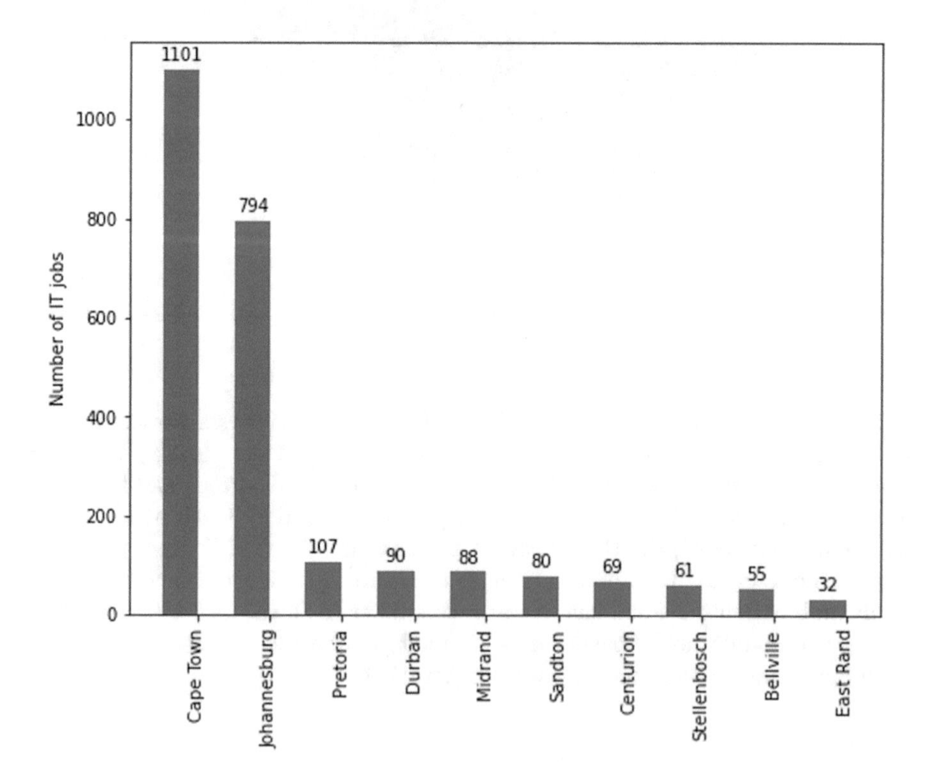

Fig. 4. The leading ten cities for IT jobs in South Africa

was second (794 jobs), followed by Pretoria (107 jobs), with 24% of all IT jobs in the country.

Three (3) of the cities mentioned in Table 2 are in the Western Cape province (Cape Town, Stellenbosch, Bellville), six (6) are in Gauteng province (Johannesburg, Pretoria, Midrand, Sandton, Centurion and East Rand) and only one is in KwaZulu-Natal (Durban).

Our finding that Cape Town is the leading city for IT jobs in South Africa is consistent with the fact that the city tech ecosystem attracted foreign direct investment of over one billion South African Rand in 2020 [16]. Cape Town has more than 450 IT companies employing over 40,000 people [17]. The city is one of the biggest digital start-up ecosystems in the country and aims to be the technology powerhouse of Africa. Many tech giants (including Naspers, Aerobotics and a subsidiary of Amazon Web Services) have their headquarters there. In terms of opportunities for individuals in the IT field, Johannesburg and Cape Town are the main centres of attraction. There are claims that people working in Johannesburg earn more (sometime, as much as 30% more) than their counterparts in other South African cities [18].

5 Study Implications

The findings of this study have practical implications for many stakeholders, including companies, governments, academic institutions, IT students and IT professionals.

IT Students and Professionals: Students (potential job seekers and entrepreneurs) and current professionals may make decisions about their future career goals based on the information presented in this article.

Government: The findings of this study, such as the fact that three of the nine provinces account for over 90% of the country's IT jobs, may be of help to policymakers in terms of resource distribution and other strategic decision-making.

Companies: Locations with the most IT jobs will have some of the best IT ecosystems and talent because of the ripple effect. As a result, our findings will be of use to firms developing IT strategies, particularly those relating to staff retention and talent recruitment.

6 Industrial Applicability of the Results

In terms of industry, recruitment companies are the biggest providers of IT jobs in South Africa and the fact that DigiOutsource (the company with the highest volume of IT jobs vacancies) is a digital outsourcing giant support the opinions expressed in the literature regarding the huge demand for jobs with flexible job locations.

Career opportunities for IT jobs with "remote-working-capabilities" such as DevOps Engineer, Software Engineer have increased significantly during the pandemic and likely to continue. Similarly, advancements in virtual conferencing tools appear to have lowered the obstacles to people connecting and expanding their professional networks. As such, the democratization of job opportunities can be expected as well as the ease of migration of skilled jobseekers across the country - and beyond.

Although our analysis shows that majority of the IT jobs in South Africa are concentrated in Cape Town and Johannesburg, companies in locations with relatively smaller talent-pool (e.g., Northern Cape, Limpopo) can now find diverse talent more easily through remote-work opportunities, especially from technical domains that are underrepresented in their area or for skills that are locally scarce. In the same vein, job seekers in locations with less job postings (or other career opportunities) may explore employment options with remote-working capabilities.

7 Summary and Conclusions

7.1 Summary

We retrieved and analysed 3,355 IT jobs posted on the LinkedIn South Africa website for this study. We were looking for answers to the following research questions (RQs) with a view to identifying the top locations for IT jobs in South Africa, and the organisations that are hiring:

RQ 1: Which IT jobs are in the highest demand? The most frequently recurring job title was DevOps Engineer, followed by Software Engineer and then Data Engineer. In terms of functions, we discovered that jobs in IT support were most in demand, followed by software development and, finally, "data" jobs. RQ 2: Which companies advertised the most job vacancies in IT? DigiOutsource was the company with the largest number of IT vacancies, followed by takealot.com (industry: Retail) and then impact.com (industry: Computer Software). However, the top three companies with the largest number of IT vacancies represented IT & Services, Retail and Computer Software, respectively. Recruitment companies posted half (50%) of the leading IT jobs in South Africa, compared with 20% from Financial Services companies.

RQ 3: Which locations (provinces and cities) had the largest number of job vacancies in IT? Three provinces accounted for 91% of all IT vacancies in South Africa, namely Gauteng (1,568 jobs, frequency: 46.7%), Western Cape (1,321, frequency: 39.4%) and KwaZulu-Natal (162, frequency: 4.8%). Job distribution based on cities showed that Cape Town (1,101 jobs) hosts 33% of all IT jobs in South Africa. In other words, one in every three IT jobs is located in Cape Town. The other two major cities in this respect were Johannesburg (794 jobs) and Pretoria (107 jobs). The province with the least number of IT vacancies was Northern Cape (12, frequency: 0.4%), followed by Free State (14, frequency: 0.42%) and Limpopo (27, frequency: 0.8%).

7.2 Conclusion

We conclude from our analysis that Cape Town is the leading location for IT jobs in South Africa, followed by Johannesburg and then Pretoria. According to

Provinces, Gauteng has the highest number of IT jobs in the country, followed by Western Cape and then KwaZulu-Natal. The province with the least number of IT vacancies was Northern Cape, followed by Free State and then Limpopo. The three most in-demand IT jobs were DevOps Engineer, Software Engineer and Data Engineer. The three companies with the largest number of job vacancies in IT were DigiOutsource, takealot.com and impact.com. Finally, it is interesting to note that majority (50%) of the leading IT jobs posted on LinkedIn in South Africa came from Recruitment companies.

7.3 Limitations and Threats to Internal Validity

First, we were limited by the required word count for this publication. Second, this study focused only on "location," which is one of the many fields in our dataset. The dataset utilised in this work was extracted between May 2021 and August 2021: if the extraction had continued over a longer time, we might have acquired a larger dataset. Our feature sets were relatively small, which produces the current analysis.

One of the threats to internal validity was that the locations specified on some of the online job postings required manual fixings. We mitigated this threat by assigning such tasks to the individual familiar with the local (South African) geography and understanding the IT industry.

7.4 Areas for Future Research

There are numerous possibilities for scholars who would like to expand on this subject. First, the required technical skills and educational qualifications for each job title could be retrieved and analysed to determine what employers are looking for in job applications. Second, future researchers will be able to extract more relevant fields from job vacancies (such as job functions, industry, income and seniority level). We also noticed that most of the top IT jobs in South Africa were advertised by recruitment firms: it would be interesting to see if this was a coincidence or if it has become a pattern in South African IT business.

References

1. Near, J.P., Rice, R.W., Hunt, R.G.: Work and extra-work correlates of life and job satisfaction. Acad. Manag. J. **21**(2), 248–264 (1978)
2. Dada, O.A., Sanusi, I.T.: The adoption of Software Engineering practices in a Scrum environment. Afr. J. Sci. Technol. Innov. Dev. **1**(1), 1–18 (2021). https://doi.org/10.1080/20421338
3. Dada, O.A., Olaleye, S.A., Sanusi, I.T., Obaido, G., Aruleba, K.: Customer satisfaction: refunds from European airlines during COVID-19. Int. J. Innov. Creativity Change **15**(10), 287–304 (2021)
4. Yunusa, A.A., Sanusi, I.T., Dada, O.A., Oyelere, S.S., Agbo, F.J.: Disruptions of academic activities in Nigeria: university lecturers' perceptions and responses to the COVID-19. In: 2020 XV Conferencia Latinoamericana de Tecnologias de Aprendizaje (LACLO), pp. 1–6. IEEE (2020)

5. Dada, O.A., Olaleye, S.A., Sanusi, I.T., Obaido, G.: COVID-19 and airline refunds: an analysis of flight passengers' reviews in North America. In: 19th International Conference e-Society, pp. 201–208 (2021)
6. Sanusi, I.T., Olaleye, S.A., Dada, O.A.: Teaching experiences during COVID-19 pandemic: narratives from ResearchGate. In: 2020 XV Conferencia Latinoamericana de Tecnologias de Aprendizaje (LACLO), pp. 1–6. IEEE (2020)
7. International Labour Organization 2020: Teleworking During the COVID-19 Pandemic and Beyond, 1st edn. International Labour Organization, Geneva, Switzerland (2020)
8. Job Location More Important to UK Workers than Wage. https://www.dovetailrecruitment.co.uk/workers-wage-location-importance. Accessed 24 Oct 2021
9. Job Location the Most Influential Factor When Accepting a Job Offer Ceridian Report Reveals. https://www.businesswire.com/news/home/20181009005066/en/Job-Location-the-Most-Influential-Factor-When-Accepting-a-Job-Offer-Ceridian-Report-Reveals. Accessed 9 Oct 2021
10. Emerging Market Insights: Is South Africa the Next. Africa's Economic Powerhouse, Johannesburg. https://mg.co.za/article/2019-02-22-00-africas-economic-powerhouse. Accessed 4 Oct 2021
11. Which is the African Powerhouse, Nigeria or South Africa? https://www.cfr.org/blog/which-african-powerhouse-nigeria-or-south-africa. Accessed 11 Nov 2021
12. South Africa Unemployment Rate Rises to Highest in the World. https://www.bloomberg.com/news/articles/2021-08-24/south-african-unemployment-rate-rises-to-highest-in-the-world. Accessed 1 Nov 2021
13. The 33 most recruited jobs and how to proactively grow your talent pipeline. http://www.linkedin.com. Accessed 21 Aug 2021
14. X5 Reasons Why DevOps Is a High-Demand Job. https://dzone.com/articles/5-reasons-why-devops-is-in-demand. Accessed 31 Aug 2021
15. About DigiOutsource. https://www.offerzen.com/companies/digioutsourc. Accessed 4 Oct 2021
16. Cape Town becomes major tech start-up investment hub. https://www.itweb.co.za/content/mQwkoM6P3x3v3r9A. Accessed 10 May 2021
17. South Africa after COVID-19 - Accelerating job creation. https://blogs.worldbank.org/africacan/south-africa-after-covid-19-accelerating-job-creation. Accessed 13 July 2021
18. Why Gauteng is by far the best province to work in. https://careeradvice.careers24.com/career-advice/money/salaries-in-sa-gauteng-is-the-highest-paying-city-by-far-20150929. Accessed 29 Sept 2021
19. X9 Future of Work Trends Post-COVID-19. https://www.gartner.com/smarterwithgartner/9-future-of-work-trends-post-covid-19. Accessed 29 Apr 2021
20. How the world of work may change forever. https://www.bbc.com/worklife/article/20201023-coronavirus-how-will-the-pandemic-change-the-way-we-work. Accessed 21 Oct 2021
21. The future of work after COVID-19. https://www.mckinsey.com/featured-insights/future-of-work/the-future-of-work-after-covid-19. Accessed 20 Feb 2021
22. CareerJunction Index - executive summary, August 2021. https://cj-marketing.s3.amazonaws.com/CJI_Executive_Summary.pdf. Accessed 1 Aug 2021

Business Model Reinvention: Impacts of Covid-19 in the Hospitality Business

Olexandr Adam[1], Carlos Jeronimo[1], Leandro Pereira[2](✉), Rui Gonçalves[1], Alvaro Dias[2], and Renato Lopes da Costa[2]

[1] ISCTE-IUL, Av das Forças Armadas, Lisbon, Portugal
{olexandr_adam,carlos.jeronimo}@iscte-iul.pt
[2] BRU-Business Research Unit, ISCTE-IUL, Av das Forças Armadas, Lisbon, Portugal
{leandro.pereira,alvaro.dias,renato.Lopes.Costa}@iscte-iul.pt

Abstract. Currently, the hospitality business is facing severe challenges due to the COVID-19 pandemic, forcing hotels to quickly adapt to a new operational reality and address in a critical way the sustainability challenge. Such adaptation requires redefining the current business models since they are not adequate for the present global scenario. The study primary goal is to analyse the main impacts of COVID-19 in the hospitality business and propose a general reinvented business model, aiming to contribute to the hospitality industry's reinvention in the global scenario. A literature review was developed to fulfil these objectives, and an in-depth interview was conducted with 14 professionals who work at different hotels. The obtained results demonstrated the reinvention of the hotels' business model must include significant investment in technology and digital communication since these two are vital for the industry's growth and evolution by providing customers with a sense of trust.

Keywords: hospitality business · business models · digital transformation · COVID-19 · pandemic · BMIs

1 Introduction

According to the World Tourism Organization (UNWTO, 2020), over the past decades, the tourism industry has been experiencing exponential growth and a deepening diversification, which justifies why this industry has become one of the fastest-growing economic sectors in the world, becoming a key driver for socio-economic progress.

This exponential growth is related to the implementation of business models, a tool that organisations use to reach and offer value to their customers, including channels, partnerships, and internal activities or resources. Amongst the multiple frameworks of business models, the most popular one, the Business Model Canvas, proposed by Osterwalder (2004), which comprises nine building blocks referring to the following aspects of a company: 1) key partners; 2) key activities; 3) key resources; 4) value proposition; 5) customer relationships; 6) channels; 7) customer segments; 8) cost structure; and 9) revenue streams (Osterwalder and Pigneur, 2010).

V. Gupta et al. (Eds.): SSEBIM 2022, LNISO 62, pp. 75–97, 2023.
https://doi.org/10.1007/978-3-031-32436-9_7

The current digital era introduces the need to upgrade and adapt the existing business models, especially by implementing digital technologies to attract more customers, maintain the pertaining industry's growth, and keep a competitive advantage (Ivan, 2020).

Now, the hospitality business is facing serious challenges due to the COVID-19 pandemic. This pandemic crisis has forced hotels to quickly adapt its operational reality, redefining processes and communication strategies to innovate in terms of the customers' experience.

This study analyses the major impacts of COVID-19 in the hospitality business – especially in hotels –, namely in terms of the necessary business model reinvention. The research also intends to identify the main changes hotels had to implement to adapt their business in the pandemic context. The goal is to present a general post-COVID-19 business model to contribute to the hospitality industry's reinvention in the global scenario. Hence, the research question explored is: *How was the business model of hotels reinvented after COVID-19?*

The qualitative technique used to collect the primary data refers to in-depth interviews, allowing reconstruction of the participants' experiences. The study sample comprises 14 interviewees working at different hotels.

2 Literature Review

2.1 Hotel Industry

The hotel industry involves all tourism entities that provide accommodation and other services for tourist consumption, being organised in various forms and companies to satisfy customers' wishes and needs while simultaneously achieving economic goals and being linked to several other economic activities. This industry is classified according to the following categories:

1) Accommodation services (hotel accommodation units);
2) Food and beverage services (bar and restaurant);
3) Sports and recreation services (pools, golf courts, gyms, and event rooms);
4) Cultural and entertainment services (concerts, exhibitions, and conferences);
5) Merchant services, souvenirs, and personal necessities;
6) Trades and services (nail salons, hairdressers, and photographers);
7) Health and other services (rehabilitation, treatment, and diagnosis (Batinić, 2016)).

Other services and operation providers complement the hotel industry, directly related to travel agencies, tourism organisations, marketing organisations, transportation services, convention centres, information services, catering organisations, among others. Therefore, the hotel industry is influenced by multiple services and companies, contributing to its success (Batinić, 2016).

2.2 Business Models

Osterwalder and Pigneur (2010:14) define the business model as the "rationale of how an organisation creates, delivers, and captures value". According to Chesbrough and Rosenbloom (2002), who present a more detailed definition, the function of a business model comprehends different aspects:

1) To demonstrate the value proposition;
2) To identify the market segmentation and the revenue generation;
3) To define the structure of the value chain;
4) To specify the revenue mechanism;
5) To assess the cost structure;
6) To clarify the position of the company in the value network;
7) And to formulate the competitive strategy.

The Business Model Canvas, proposed by Osterwalder (2004) and later improved by Osterwalder and Pigneur (2010), who suggested a more practical business model framework (Fig. 1) that allows companies to develop and change business models, and therefore to create, deliver and capture value for their customers.

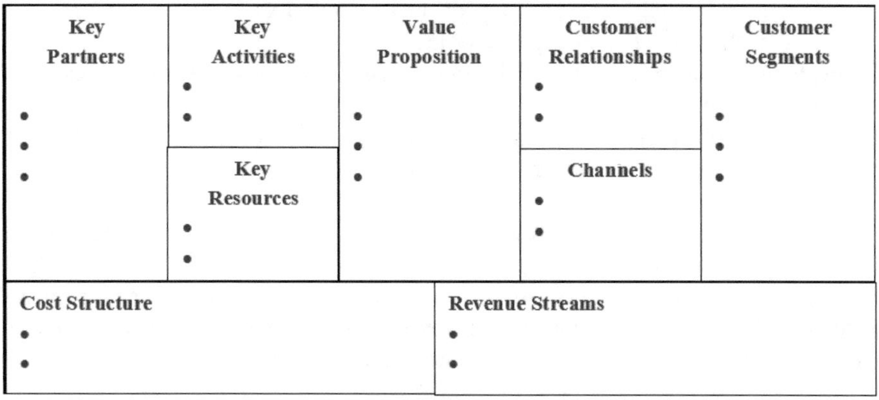

Key Partners	Key Activities	Value Proposition	Customer Relationships	Customer Segments
• • •	• •	• • •	• •	• • •
	Key Resources		Channels	
	• •		• •	
Cost Structure • •			Revenue Streams • •	

Fig. 1. The Business Model Canvas proposed by Osterwalder and Pigneur (2010)

2.2.1 Business Model Innovations vs. Reinventing Business Model

According to Mitchell and Coles (2003: 17) Business Model Innovations (BMIs) refer to "the replacements within the business models that companies make to provide product or service offerings to customers and end-users that were not previously available". Different business models compete within the same markets, while the boundaries between the existing industries are blurring. Hence, these authors consider that BMIs are not about benchmarking or copying competitors but about creating new mechanisms that aim to create value and derive revenues Osterwalder and Pigneur (2010).

Johnson (2010) claims that BMIs can reshape industries and redistribute value. BMIs are crucial in times of instability, given their ability to deliver superior returns by defending and protecting a dying core business or exploring new avenues for growth. Nowadays, technological innovations alone do not ensure high profits for companies, being also necessary to implement an innovative business model for companies to reap advantages from such technological innovations (Lindgardt et al., 2009).

Mitchell and Coles (2003) emphasise that improving the business model will make it difficult for competitors to follow the company's actions and copy them, creating a competitive advantage. There are some conditions that business models must fulfil to provide companies with a sustainable competitive advantage, more precisely:

1) The business model must meet specific customer needs;
2) It must be a non-imitable model in some aspects;
3) The company needs to be quite flexible to respond to eventual changes in the business model (Teece, 2010).

The reinvention of business models is solely related to creating a new business model, based on four dimensions.

1) Customer sense: ease of acceptance of a new value proposition;
2) Technology sensing: strength, impact, and direction of technology on the new customer value and the business network;
3) Economic/profitability sensing: business model's economic feasibility and profitability;
4) Business infrastructure sensing: traditional business network's responsiveness to reconfigure the new business model (Voelpel et al., 2004).

These four dimensions are interactive with each other. Adequate technology must be in place within the company, aiming to leverage efficiency and the new customer value. In turn, the business system infrastructure must be configured to enhance the created customer value. All the undertaken endeavours must be economically feasible to truly benefit those involved in the entire reinvention process (Voelpel et al., 2004).

2.2.2 Business Model for the Hotel Industry Definition

Considering that hotels operate in unique internal and external contexts, it is important to understand and follow the strategic management practices. The unique characteristics of hotels are essentially related to the following aspects: 1) inseparability of customer participation in the service process; 2) perishability; 3) simultaneity; 4) intangibility; 5) cost structure; 6) heterogeneity; and 7) labour intensity (Okumus et al., 2019).

The hotels' cost structure influences the companies' managerial and resource allocation, affecting the strategy formulation process, which frequently involves sustainable-oriented strategies and pricing strategies to maintain their profit. Also, hotels are very labour intensive, hence requiring a significant number of employees. Therefore, hotels can only take care of their customers if they first take care of their employees, since the latter will strongly participate in delivering a unique experience for customers (Okumus et al., 2019).

A business model provides the hotel's method for making money in the current business environment. It provides a deeper insight into the alignment in terms of strategies and actions while supporting strategic competitiveness. Table 1 presents a general business model canvas for the hotel industry, which is based on the nine blocks proposed by Osterwalder and Pigneur (2010) (Table 2).

Table 1. General business model canvas for the hotel industry

<table>
<tr>
<td rowspan="2">Key Partners
- Travel and tourism agencies
-Airline companies
- Franchises
- Owners
- 3rd party booking sites
- Travel agencies
- Corporate partners</td>
<td>Key Activities
- Hospitality
- Hotel management</td>
<td rowspan="2">Value Proposition
- A uniform travel experience customers can trust</td>
<td>Customer Relationships
- Direct relationship with customers
- Indirect relationship with customers (through 3rd party internet sites)</td>
<td rowspan="2">Customer Segments
- Leisure travellers
- Business travellers
- Corporate customers
- One-off events</td>
</tr>
<tr>
<td>Key Resources
- Customer relationships
- Hospitality industry expertise
- Timeshares</td>
<td>Channels
- Online 3rd party booking sites
- Online booking channels
- Travel agents</td>
</tr>
<tr>
<td colspan="3">Cost Structure
- Labor costs
- Fixed assets costs
- Depreciation
- Management costs
- Renovation costs
- 3rd party partner fees</td>
<td colspan="2">Revenue Streams
- Assets renting (parking for example)
- Customer sales
- Business contracts</td>
</tr>
</table>

Source: Elaborated by the author, adapted from https://miro.medium.com/max/1154/1*R04WPudk3YuTjosTszjdYw.png

2.3 Evolution of the Digital Transformation in Pre-COVID-19 Scenario

Currently, the hotel industry needs to invest in technology properly and fully understand how to exploit it, aiming to meet the challenging emerging trends. The greatest challenge

Table 2. Main impacts of COVID-19 in the three main stakeholders in different stages (Sigala, 2020)

Stakeholder	Respond stage	Recovery stage	Restart stage
Tourism demand	– Trip cancellations – Loss of money paid for travel-tourism – Quarantine and social distancing – Travel restrictions and bans – Panic buying and stockpiling	– Social distancing – Lockdown and stay at home – Use of technology for contactless services (shopping, working, and studying) – Virtual events and communication	– New tourism service and experience based on digital technologies – New priorities determining the tourists' selection, evaluation, and consumption behaviour
Tourism operators/businesses	– Managing the safety and health of tourists and employees – Handling customer communication (changing travel bookings and itineraries, cancellation of bookings, and compensations and refunds)	– Ensure business continuity and building resilience (repurpose of resources, innovation from necessity, acceleration of digital adoption, customer and employee engagement, and mitigate crisis impacts)	– Resetting the new business normal (re-opening, new cleaning and hygiene protocols, crowd management, social distancing practices, and re-design and re-imagine the customer journey to make it contactless)
Destinations and policymakers	– Ensuring the health and safety of tourists – Crisis communication – Managing repatriation of citizens – Interventions to support vulnerable employees/tourism businesses	– Keeping tourists informed and interested – Virtual visits to destinations – Engaging with destination partners and stakeholders – Crisis communication – Interventions to support the tourism industry and jobs	– Reimagine new types of sustainable and responsible tourism – Setting safety and health regulations – Develop strategies for staged re-opening – Promotion and motivation to tourists – Health passports and identities

that the industry faces is related to the determination of both the type and the level of technology that it must integrate into their facilities. The emerging technologies that have been evolving and gradually implemented in the hotel industry are the following ones:

1) Self-service technology;
2) Mobile devices;
3) Artificial intelligence;
4) Smart mirrors;
5) Smart-home assistants;
6) Robots (Mairinger, 2018).

The self-service technologies allow customers to be served without establishing direct contact with the hotel's employees, consisting of a more productive approach. The most common type of self-service technology refers to kiosks systems in hotels, which allow customers to check-in and check-out on their own (Wei et al., 2017). Mobile devices, on the other hand, have been constantly used in hotels as room-keys and credit cards, with their cameras reporting who is entering and leaving with the assistance of recording software. These mobile devices use help to increase security and provide a more personalised service to customers, which justifies their significant benefit when implemented in the hotel industry (Tomašević, 2018).

Artificial intelligence is frequently used to mine data from customers, namely from external sources, to create a more advanced customer profile and, ultimately, a more personalised service. Regarding smart mirrors, they can be introduced in the hotel rooms to provide several types of information to customers, namely related to the weather, their emails, the stock market, and even interact with the customers' smartphones. Similarly, smart-home assistants can provide real-time connectivity and convenience to customers, who can request assistance or service in a much easier and faster manner (Hospitality Technology, 2018).

Lastly, robots are still an upcoming technology, even though some hotels already have implemented them in several roles, such as receptionists, housekeeping, bartenders, and concierges. Robots can move at the same speed as people and are based on a technology platform for cognitive computers, combining several application interfaces. However, the implementation of robots in hotels must be carefully assessed. Human interaction is considered an essential aspect of customers' entire hotel experience (Mairinger, 2018).

2.4 COVID-19 Crisis in the Hotel Business

2.4.1 Impacts of COVID-19

The COVID-19 was declared a pandemic in March of 2020 and significantly impacted the global economic, socio-cultural, and political systems. Several health communication measures and strategies, such as community lockdowns, social distancing, self- or mandatory-quarantine, and stay at home campaigns, have ceased global tourism, travels and leisure. Although the tourism industry is used to bounce back from distinct types of crises/outbreaks, the COVID-19 is considered to be a different type of crisis, given the fact that it can have long-term and profound structural and transformational changes within the tourism sector, leading to a worldwide recession (Novelli et al., 2018).

The COVID-19 impacts on tourism are estimated to be very harmful worldwide, since the international tourist arrivals are estimated to drop about 78%, referring to a loss of US$ 1.2 trillion in export revenues from tourism and 120 million tourism job cuts.

Hence, the current discussions and research regarding the tourism industry aim to use the COVID-19 pandemic as a transformative opportunity (Sneader and Singhal, 2020).

One of the main consequences of COVID-19 is related to both the necessity and the acceleration of the use of technologies to help tourists (e.g., information regarding travel restrictions, hygiene measures, online crisis communication, and online COVID-19 alerts) and businesses (e.g., festivals/events, online food delivery, destinations, and virtual visits of museums). Still, not everyone has access to technology, which undermines the assumption that everyone can effectively use technology tools and the conveyed information. Digital division found between consumers and businesses has converted COVID-19 into an infodemic and a tool that deepens the economic division and the competitive gap between tourism operators of different sizes Sigala (2020).

According to Sigala (2020), COVID-19 has several impacts and implications on the three main stakeholders in the tourism industry (the tourism demand, the tourism operators, and the destinations and policymakers), namely in three different stages, which refer to the response, the recovery, and the restart stage from the current pandemic.

However, the current pandemic of COVID-19 also impacts business models, including those referring to the tourism industry. Ritter and Pedersen (2020) concluded that there are six types of crisis impacts, which essentially describe the different ways a crisis can profoundly influence a business model. These six impact types refer to the following business models: 1) Antifragile; 2) Robust; 3) Adaptive; 4) Suspended; 5) Aided; 6) Retired.

Antifragile business models present a better performance during a crisis, given the fact that they tend to improve under stress and better realise their potential during that scenario. Certain previously considered costly capabilities, unproductive, and complex, became important once the crisis emerged. In these business models, the main managerial issues refer to detecting the favourable environment and accelerating the business model itself (Ritter and Pedersen, 2020).

Robust business models are known because they remain in the desired operational state, even during a crisis, only suffering some changes in terms of the business model's volume, but not in the business model in itself. The main managerial issues are related to the insurance of the business continuation and the management of the changes regarding the volume. Adaptive business models refer to models that need to change when a crisis hits, for instance, regarding urgent healthcare needs (e.g., using masks). Moreover, these changes might include shifting from on-site services to remote services to continue operating and selling their products/services. Hence, the main managerial issues within this type of business models include identifying the opportunities to develop the model and the rapid implementation of those identified initiatives (Ritter and Pedersen, 2020).

Suspended business models are related to a temporary closure of a business model, to reopen once the crisis is over. The major key assumptions for these specific business models are the following: 1) the business model will be running profitably once the crisis is over; and 2) it is preferable to invest during the suspension period rather than to dissolve the business and start a new venture after the crisis is over. Thus, the managerial issues that arise with suspended business models are associated with the insurance of enough capital to finance the business during its closure period and the insurance of the access to resources, namely human resources, after the business restarts (Ritter and Pedersen, 2020).

Aided business models are related to businesses that cannot support themselves during a crisis, therefore depending on external support. This external support might be provided by governments, investors, or management teams, with the main goal of supporting businesses that are quite vulnerable during a crisis. The main assumption regarding these business models is related to the fact that once the crisis is over, businesses will be successful once again, being preferable to rely on external aids rather than closing their ventures and/or firing everyone due to lack of profit. The main managerial issues in these models include the presentation of the business model as being worthy of receiving external aid, the identification and further application of the aid, and the preservation of the business model for a later revival, especially after the crisis is over (Ritter and Pedersen, 2020).

Retired business models are those that cease to exist during a crisis, since the cost of maintaining the business model operating throughout the crisis exceeds the expected profits after the end of the crisis. In this case, it is preferable to end and close the business and to start a new one after the crisis is finally over. The managerial issue of retired business models refers to the organisation of an orderly shutdown and exit of the business (Ritter and Pedersen, 2020).

Sustainability continues as one of the most commonly discussed trends within the hotel industry (Kim and Barber, 2019). Sustainability, as a concept that contains environmental, economic, and socio-cultural dimensions, and is most commonly defined as "development which meets the needs of current generations without compromising the ability of future generations to meet their own needs" (United Nation Economic Commission for Europe, 2017). Before implementing any specific operational practices or certifications to deal with COVID-19, the hospitality industry should consider how it can be integrated into a strategic framework. The industry needs to determine which SDGs to focus on to fulfill their sustainability objectives and align with their business. Jones and Comfort (2020) say that any of the leading players within the hospitality industry have emphasized their commitment to the United Nations Sustainable Development Goals (SDGs), designed to achieve a global transition to a more sustainable and resilient future, however, there are fears that without financial resources the COVID-19 crisis will change the priorities.

2.4.2 Digital Transformation Impacts on Business Model Redefinition

Tourists' concerns regarding health risks and security are increasing, due to the current COVID-19 pandemic, affecting travel behaviours. Hotels must address all these changes and restore travellers' sense of security. There are three key dimensions to consider in terms of management: 1) artificial intelligence; 2) hygiene and cleanliness; and 3) health and healthcare (Jiang and Wen, 2020).

Firstly, hotels are starting to pay attention to artificial intelligence's potential benefits and their applications in hotel management practices. This might be immensely helpful to hotels, given the role of social distancing as an effective prevention strategy against COVID-19. Indeed, by adopting artificial intelligence, hotels can help to protect their customers and their employees, increasing the general sense of security and trust.

Lukanova and Ilieva (2019) present some examples, in the lobby, several electronic podiums welcome guests for check-in, making the first analysis of faces and ID cards. In the elevators, faces are recognized again to verify the floors of the building the customer can access. Secondly, hygiene and cleanliness are essential in the current scenario, given that they have been a recurring issue in pandemic outbreaks. Thus, these two dimensions are critical to hotels since they increase customers' sense of security and safety, but increase costs too with more products and labor work for clean. Lastly, health and healthcare are also especially important dimensions to reassure customers and to increase their trust in the hotel (Jiang and Wen, 2020).

Mastrogiacomo (2020) proposes ten specific digital strategies that the hotel industry must consider, especially if hotels want to surpass the current crisis and thrive. According to the author, hotel managers should ponder on these ten considerations when preparing their digital strategy:

1) Consistently align the marketing and revenue management strategy;
2) Consistent fresh content on the hotel's website, since customers will expect up to date content;
3) Maintain a flexible cancellation policy;
4) Delicately balance the direct and Online Travel Agency strategy;
5) Develop a month-to-month marketing plan;
6) Maximise website conversions;
7) Stay connected to guests who have cancelled reservations;
8) Revisit the automated marketing campaign strategy;
9) Maximise revenue by focusing on upselling ancillary services (experiential packages with several services included);
10) Get creative with the hotel inventory, such as providing co-working spaces and meeting rooms.

3 Methodology

3.1 Research Objectives and Questions

Based on the main research objectives present in Table 3, the key research question explored is: *How was the business model of hotels reinvented after COVID-19?*

To answer to this central research question, several sub research questions were developed with the main goal of supporting the present study. Table 3 summarises the connection between these three elements: research objectives, sub research questions, and literature review.

3.2 Research Approach

Considering both the research objectives and questions, it is ultimately necessary to understand the opinions, behaviours, and interactions of people that are, or were, involved

Table 3. Sub research questions of the study and intrinsic research objectives and literature review (Source: Elaborated by the author)

Research objectives	Sub research questions	Literature review
i. To analyse the influence that COVID-19 has in the hospitality business, especially in hotels	Did reservations decrease? Was the staff fired or put in lay-off? Did the services suffer any changes or were cut back?	Sigala (2020)
ii. To identify the main changes that hotels must implement to adapt their businesses to the current context of COVID-19	Were there any changes in terms of the hotel's management and operations? Did all the operations and services suffer severe changes?	Sigala (2020) Jiang & Wen (2020)
iii. To reinvent/present a general business model that can benefit the hospitality industry in the current global scenario	What changes must be implemented in the business model of hotels? What new dimensions must be considered in the business model?	Mastrogiacomo (2020) Mairinger (2018) Jiang & Wen (2020)

with hotels. Therefore, qualitative research was used to represent the views and perspectives about the participants' real-life events in the study (Yin, 2011) and their context (Creswell, 2007).

The present study does not comprehend an in-field observation and collection of data, mainly due to the restrictions currently imposed by the COVID-19 pandemic. The qualitative technique used to collect the primary data refers to in-depth interviews, with the main goal of reconstructing the participants' experiences and reality.

3.3 Data Collection Procedures

The study sample is composed of 14 interviewees, the interviewees were selected according to the researcher's network contacts, who had been working in the pertaining hotel both before and after the COVID-19 and considering their experience and expertise regarding the changes that occurred in the hotel due to COVID-19. In terms of geography, the hotels where the interviewees' work are mostly located in Portugal (9), while the remaining (5) are located in other countries. This disparity is beneficial to the research since it contributes to the understanding of how foreign hotels are dealing with COVID-19 and with its impacts in terms of the management of their businesses. All interviewees work daily at the pertaining hotel, and none of the hotels comprehended in the research is still closed.

Concerning the specific methods, all interviews were conducted by phone call, with the main topics being pointed out by the researcher throughout the conversation. The interviews' average duration was about 30 min and occurred during September of 2020, corresponding to a period of 6 months of changes after the spread of COVID-19.

3.3.1 Sample Characterisation

The 14 interviewed participants in the present research, all referring to different hotels with different characteristics (rooms, prices, and rankings) and different functions. All interviewees provided information according to their knowledge and experience as workers at the pertaining hotel, which essentially represents some of the biggest changes implemented in the current context, either nationally or internationally.

It is noteworthy to mention that all the hotels present a different approach to the current pandemic, implementing distinct measures and changes according to their resources and amenities, which positively contributes to the pertaining scientific field, especially by providing multiple strategies and measures that can be adequately adapted to every single hotel.

The interviewees are characterised by the hotel where they work and the functions they develop as employees of that same hotel. Overall, the participants of this research are receptionists (4), food and beverage (F&B) supervisors (3), guest relations (2), chefs (1), event coordinators (1), front desk agents (1), front office managers (1), and sales coordinators (1), as it is demonstrated and summarised in Table 4.

In terms of the hotels' occupation, the number of rooms that each one provides diverges significantly, ranging from 9 up to 500 rooms. This information is rather important since it characterises the hotels' dimension, both in terms of occupation and business. Most of the hotels have between 100 and 200 rooms (7 hotels), while the remaining have either less than 100 rooms (5) and about 200–500 rooms (2). Another important aspect that must be analysed to characterise the comprised hotels is the average price per night that each hotel offers to the public. The hotels' price range per night varies significantly, with the vast majority of the hotels charging 100–150€ per night (6), the remaining ones charge 150–200€ (4), 50–100€ (2), and 250–300€ (2).

From the analysis of the portrayed graphs, in a general sense, it is possible to conclude that the majority of all interviewees work as receptionists (4) in the pertaining hotel, most of them located in Portugal (9), having an occupation of about 100–200 rooms (7) and charging between 100–150€ (6) per night. It is noteworthy to add that all hotels present a remarkably similar ranking, only varying between 5 and 4 stars, making them very prestigious and luxurious hotels.

4 Data Analysis

This chapter's main objective is to gather information regarding the current scenario in the hotel industry and present a thorough and precise business model, illustrating the hotel's strategy both before and after such pandemic crisis.

The analysis of the obtained results is divided into different subchapters, referring to one of the nine blocks of the Business Model Canvas proposed by Osterwalder and Pigneur (2010). However, in the following paragraphs, the interview's introductory questions will be analysed, referring to open questions asked to interviewee. The obtained results will be confronted with the literature review.

The first introductory question was: *After the opening of hotels post-COVID, what are the* ***main changes*** *that you observed? What has changed in hotels?* According to the collected answers, the main changes that occurred in hotels after COVID-19 are related

Table 4. Characterization of the sample of the study (Source: Elaborated by the author)

Interviewees	Hotel description	Function
1	Hotel 1–76 rooms, 140€/night, 5 star-hotel, Oeiras	Receptionist
2	Hotel 2–26 rooms, 120€/night, guest house, Lisbon	Chef
3	Hotel 3–194 rooms, 170€/night, 5 star-hotel, Lisbon	F&B Supervisor
4	Hotel 4–186 rooms, 200€/night, 5 star-hotel, Lisbon	F&B Supervisor
5	Hotel 5–9 rooms, 80€/night, guest house, Peniche	Receptionist
6	Hotel 6–177 rooms, 170€/night, 5 star-hotel, Amoreira	Front Office Manager
7	Hotel 7–30 rooms, 200€/night, 5 star-hotel, Sintra	Events Coordinator
8	Hotel 8–200 rooms, 300€/night, 5 star-hotel, Budapest	Front Desk Agent
9	Hotel 9–194 rooms, 300€/night, 5 star-hotel, Dubai	Guest Relations
10	Hotel 10–192 rooms, 120€/night, 5 star-hotel, Cascais	Guest Relations
11	Hotel 11–283 rooms, 130€/night, 5 star-hotel, Abu Dhabi	Sales Coordinator
12	Hotel 12–500 rooms, 120€/night, 4 star-hotel, London	Receptionist
13	Hotel 13–40 rooms, 50€/night, 4 star-hotel, Óbidos	Receptionist
14	Hotel 14–140 rooms, 130€/night, 4 star-hotel, UK	F&B Supervisor

to new safety measures (topic identified by 11 interviewees), directly associated with the safety and hygiene norms disseminated by the government and by health institutions, and to limited services (11), in the sense of reducing both the staff and customers on the same space to avoid the risk of infection by COVID-19. Some other interviewees claimed that some of the changes that occurred in their hotel are associated with the reduction of the staff (7), to the decrease of reservations by customers (4), to the staff being on lay-off or even fired (4), to lower prices (3), and the reduction of costs (1).

In sum, the main changes are all interconnected, except for implementing new safety measures, considering that the decrease in reservations by customers reduces the hotel's profit, negatively impacting its staff (in lay-off or fired). Hotels are trying to lower their prices, aiming to return to their profit margin gradually. Essentially, these results are congruent with the arguments presented by Sigala (2020) and Jiang and Wen (2020) since they corroborate the severe impacts that COVID-19 has been having on hotels and the increasing focus of hotels in the health and healthcare domain, as well as in cleanliness requirements.

The second introductory question was: *What do you think will still change in hotels?* OR *Are you planning to change anything at your hotel?* The interviewees demonstrate their belief in future changes, which include some discounts to attract more customers (4), investing in more personalised services (3), implementing safety measures to guarantee the customers' safety and well-being (2), sending more staff to lay-off due to the low income and to the constraints implemented by the government and health institutions regarding social distancing (2), reducing the hotel's costs whenever possible and necessary (1), and investing in social media to keep close contact with customers and to promote the hotel (1). All of these answers corroborate the study conducted by Jiang and Wen (2020), namely in terms of the new reality regarding the constant cares in the cleaning department, and the arguments presented by Mastrogiacomo (2020), given a clear bet on social media with the main goal of maintaining closer contact with customers and to keep all marketing efforts to attract more clients to the hotel.

The third and last introductory question was: *What do you think that got better? And what got worse?* The positive impacts that the current pandemic had on hotels are related to the offering of personalised services to customers (5), to better safety measures (3), better cleaning methods (2), and better partnerships (1). On the other hand, the negative impacts are mainly related to the fact that hotels now have less staff (7), several staff members are in lay-off (3), and there are worse hotel standards (3), mainly related to the fact that it is more difficult to provide unique packages and experiences to customers to ultimately distinguish the hotel as a higher ranking one (with 4 or 5 stars).

4.1 Key Partners

Regarding the hotels' key partners, interviewees were asked whether they changed their partners and/or suppliers. Most hotels changed their suppliers (5) to decrease the hotels' costs while maintaining their service quality. However, some preferred to maintain their older suppliers (4). In terms of tourism agencies, hotels have maintained the same partners (3), aiming to strengthen such partnership as a strategy to increase their reservations and profit. On the other hand, cleaning services seem to be a quite controversial partnership, given the fact that some hotels established new partnerships (3), while others opted for internal operations (3). This was also the case in terms of the catering services and outsourcing, considering that most hotels (3) opted to develop such operations internally.

The obtained data supports the investigations conducted by Sigala (2020) and Jiang and Wen (2020), given the fact that they illustrate some of the main impacts of COVID-19 in hotels, namely related to the decrease of profit, which directly affects the contracts established with some of the hotels' key partners. Data also emphasises the clear investment of hotels in their cleaning services, which are now provided by new partnerships or done internally.

4.2 Key Activities

With regards to the hotels' key activities, interviewees were asked two different questions. The first question was: *Has anything changed in the key activities that were sold by the hotel? (Ex.: the restaurant was closed, rooms, group services, gym, and swimming pool).* As shown in Table 5, interviewees' answers pointed out some changes in 8 different activities.

Table 5. Changes in the key activities of hotels (Source: Elaborated by the author)

Activities	Changes	Interviewees
Swimming pool	Open to the general public and not exclusively to guests	1, 2
	Limited number of clients	3, 4, 7, 12, 14
Group services	All cancelled, except for a few weddings	1, 6, 8, 9, 13
	Conference meetings were reduced	6, 8, 9
Safety measures	Staff must fulfil all the safety standards (masks and disinfectants)	1, 2
Restaurant	Closed	2, 12
	Exterior space was open	2
	Limited number of clients	3, 4, 7, 8, 10, 13
	Take-away	9, 12
	Opening a new restaurant	11
Spa	Limited number of clients	3, 7
	Closed	9, 12
Gym	Limited number of clients	4, 9
Gardens	Open to the general public	7
Bar	Changed its functions	8
	Open	12

Overall, all of these changes in terms of the hotels' key activities are related to the negative impacts mentioned by Sigala (2020), which highlight that several services had to cease their activities and that several sectors of the industry had to either close or to introduce major changes to operate according to the new safety and hygiene recommendations.

The second question regarding the hotels' key activities was: *Has anything changed in terms of the hotel's management and operations? (Ex.: restructuring, personnel training, check-in processes, more thorough cleaning).* Based on the interviewees' answers, which are summarised in Table 6, the main management and operations key activities that changed the most are related to cleaning and safety services, training, restructuring/firing, and meetings.

In sum, the obtained data in this specific question are all congruent with the main investigations analysed, more precisely with those conducted by Mairinger (2018), Jiang and Wen (2020), Mastrogiacomo (2020), and Sigala (2020), since all answers gather some recommendations and insights provided by all of these authors.

Table 6. Changes in the key activities of hotels (management and operations) (Source: Elaborated by the author)

Activities	Changes	Interviewees
Cleaning and safety services	Person responsible for checking the cleaning and safety services and the fulfilment of standards	1, 2, 3, 5, 7, 10, 11
	Avoiding sharing objects with staff and clients	3, 7, 9, 10, 11
Training	Training regarding the best practices to deal with COVID-19	3, 4, 8, 12, 13, 14
Restructuring/firing	Some staff was fired	4, 6, 14
Meetings	Meetings were conducted via the internet and by emails	6

4.3 Key Resources

The new resources that hotels had to purchase after the emergence of COVID-19, in addition to masks and disinfectants, are related to technology. Hotels had to purchase appointment lists (7), with the main objective of respecting the social distancing implemented by health institutions, smartphones apps (6), aiming to provide better and faster services to customers, which also contributes to a more personalised experience, implement QR codes (5), to avoid any material resources that can transmit COVID-19, implement contactless machines (4), with the same goal of avoiding close contact between the staff and customers, implement mobile check-in (2), also making the process way easier and safe, and purchase of thermometers (3) to check customers and staff's temperature, which consists in one of the main measures that other industries are implementing to avoid further contamination.

These answers are very similar to the arguments presented by Mairinger (2018) since the author points out some of the main technologies that are being implemented in the hotel industry, emphasising that the future is definitely towards constant innovation in this specific dimension. The only difference is that now technology is being implemented not only to improve customers' experiences but also to guarantee their well-being and safety.

Regarding the hotels' accreditation by the Clean and Safe seal, which is considered an essential resource nowadays, due to the trust and confidence that it creates on customers in the current pandemic, the vast majority got accredited by the Clean and Safe seal (10). The remaining (4) are either working towards that accreditation for future profit or already have its cleaning seal.

Lastly, some interviewees suggested that the hotels' loyalty program has remained the same (5), while others claimed that it did change after COVID-19 (5). The main changes refer to the decrease in the hotels' prices and lower stays and the accumulation of points, which can be used in future reservations. The four remaining interviewees did not answer the question.

4.4 Value Proposition

In terms of the value proposition, interviewees were asked the following question: *Were there any changes that added more value to the customers or to the hotel? (Ex.: Offer of masks and disinfectants or products for individual use).* The most popular answer among interviewees refers to offering a more personalised service (5), which is directly linked to previous answers since it includes the implementation of technology to improve the customers' experience and increase the hotels' profit. However, interviewees also mentioned the use of outdoor spaces (3) to provide some of their own services, the use of smartphone apps (2), the offering of free products to customers (1), and some discounts (1) when making their reservations. Once again, these findings are congruent with the arguments presented by Mairinger (2018), since it is quite notorious the investment in new technologies to attract more customers and to maintain some of the older ones.

4.5 Customer Relationship

One of the questions regarding the customer relationship was: *Did you try to show your clients that the hotel is safe in terms of sanitation? (Ex.: sticker on the bedroom door when it was cleaned, inform that all devices were disinfected).* Most interviewees answered this question positively (8), meaning that hotels are trying to demonstrate that their facilities do follow rigorous cleaning and hygiene measures to guarantee customers' safety and well-being.

Another question about the customer relationship was the hotels' activities that became more digital after COVID-19. The only activities that truly became more digital are related to the hotels' apps (5), which were significantly improved to fully provide a unique experience to customers and the main services that the hotel offers. This trend is highly emphasised by Mastrogiacomo (2020), who firmly claims that the hotels' growth and expansion will be based on technology investment, with several applications and benefits to both the hotels and their customers.

An important aspect of the customer relationship relates to the cancellation policy of hotels. Therefore, interviewees were asked if there were any changes in terms of such specific policies in their hotels. The cancellation policy changed in almost every hotel, except in one (Interviewee 4). Overall, hotels introduced the following changes:

1) Hotels allowed customers to cancel their reservations up to 24h/48h/72h before, without further costs (5);
2) Reservations became refundable and/or the money could be used for future reservations (4);
3) Hotels offered vouchers (2);
4) Hotels avoided any cancellations, proposing a reschedule instead (1).

Once again, the obtained data confirms the arguments provided by Mastrogiacomo (2020), who also establishes that hotels must be more flexible in cancellations, especially in the current scenario. The main goal is to promote further reservations and attract customers to the pertaining hotel, offering them some flexible options in traveling.

The last question about the customer relationship related to any offerings that hotels are currently providing to their customers, such as a free night when they stay for X days or last-minute deals. Only three of the 14 hotels provide these types of offerings to their customers, interviewees said that their hotels are offering a free week for companies that work at their home offices, team building for small groups, one free night when customers book two nights, and cheaper reservations for customers who have memberships (Interviewee 3, 5 and 6).

4.6 Channels

Regarding the hotels' channels, interviewees were asked to answer two different questions: 1) *Did the hotel bet on other sources (such as the official website)?* And 2) *Have you (the hotel) reduced the travel agents?* In terms of the sources of customers' reservations, hotels have strongly maintained their traditional sources (6). Nevertheless, some hotels trusted on other websites (4), other apps (2), and other partnerships (1). Hotels might have selected these different sources to reach more target audiences and increase their scope. This joint action with other websites, apps, or partnerships will result in an increased profit for both parties while presenting more choices to eventual customers.

Specifically focusing on the hotels' travel agents, the vast majority did not reduce any of the existing partnerships, as eight interviewees have said it. Still, three hotels reduced their partnerships with travel agents, which might be related to their strategy to decrease the hotels' costs. These answers are supported by Mastrogiacomo (2020), who suggests the use of new channels to attract more customers and to increase the hotels' scope within the digital world.

4.7 Customer Segments

The customer segments are related to the main clients that hotels typically have, referring to a particular profile. About this aspect, ten interviewees claimed that the hotels' customer segments changed after COVID-19, referring to completely different customer segments. The major change among the hotels refers to more national customers and longer stays (5) and longer stays during summer (2). These results are due to the current pandemic since most frontiers had severe constraints when traveling. Hence, it is quite normal that hotels now have more national customers and longer stays since there are not many options for tourism and/or travelling. These findings are congruent with the study conducted by Sigala (2020), who describes the main impacts of COVID-19, clearly related to the decrease of reservations and travels by consumers.

4.8 Cost Structure

The first question that interviewees were asked to answer was: *Was there any increase or reduction in the hotel costs during this period? Have you tried to reduce costs in some way?* The major decrease in the hotels' costs relates to their services (9) since they had to limit, or even close, the provided services due to the social distancing measures. Still, some other aspects were considered to reduce hotels' costs, mainly their staff, and their offer gits (newspapers, magazines, and flowers). However, eight hotels increased their costs with new products, such as cleaning and disinfection products, masks, aprons, and disposable gloves for housekeeping, acrylics, dispensers, distance signalling, to improve the sanitising conditions and hotel's appeal to eventual clients. Two of the hotels introduced a new product, an adapted restaurant menu to reduce costs and waste. These results are essentially not congruent with the arguments provided by Mastrogiacomo (2020) since they do not demonstrate a clear investment in technology and in digital solutions to overcome this current crisis.

4.9 Revenue Streams

In terms of the revenue streams of hotels, interviewees were asked to answer two different questions. The first question was related to the hotels' prices, namely to their decrease or increase after COVID-19. Most hotels decreased their prices (7), aiming to attract more national customers to their facilities since the number of foreign customers truly reduced. Connected to the ambition of the hotels to increase their profits and overcome such crisis.

The obtained data confirms the arguments presented by Sigala (2020), namely regarding the negative impacts of COVID-19 in the hotel industry. The author has also stated that the hotel industry has suffered a severe decrease in terms of reservations. Most hotels' main strategy was based on decreasing their prices, aiming to attract customers and start making profit, trying to provide newer services, especially focused on the implementation of new technology in their facilities.

It is possible to conclude that the hotels' reservations significantly decreased after the emergence of COVID-19, resulting in several staff members being fired or put on lay-off. Moreover, the hotels' services suffered a few changes, either by significantly limiting the number of customers or by being completely closed. Considering the hotels' management and operations, they started to focus even more on cleaning and safety services and training their staff members with regards to the best practices to deal with COVID-19. The main changes that must be implemented in the business model of hotels refer to measures related to the current norms in terms of health and safety, with new cleaning and health services, to new and cheaper partnerships/suppliers, to invest in technology and digital communication strategies, and to the implementation of technology to benefit both the hotel and its customers, while simultaneously attracting new customers to increase its profit.

Table 7 presents a proposal of a reinvented business model for hotels. The new measures suggested by the obtained data are presented in green, while the older measures are presented in black. This colour scheme is used to make it easier to understand the measures adopted before the emergence of COVID-19 and the new measures suggested in this study for future implementation and after COVID-19. Therefore, it is expected that this reinvented business model can help hotels in the near future, even though it might not fully result in their growth and evolution.

Table 7. Proposal of a reinvented business model for hotels (Source: Elaborated by the author)

<table>
<tr>
<td rowspan="2">Key Partners
- Travel and tourism agencies
- Airline companies
- Franchises
- Owners
- 3rd party booking sites
- Travel agencies
- Corporate partners
- Cleaning services
- Technology companies</td>
<td>Key Activities
- Hospitality
- Hotel management
- Healthcare
- Restaurants and bars (take-away)
- Training (staff)</td>
<td rowspan="2" colspan="2">Value Proposition
- A uniform travel experience customers can trust
- Cleanliness and safe facilities
- Personalized services based on digital communication and apps</td>
<td>Customer Relationships
- Direct relationship with customers
- Indirect relationship with customers (through 3rd party internet sites and apps)</td>
<td rowspan="2">Customer Segments
- Leisure travellers
- Business travellers
- Corporate customers
- One-off events</td>
</tr>
<tr>
<td>Key Resources
- Customer relationships
- Hospitality industry expertise
- Technology</td>
<td>Channels
- Online 3rd party booking sites
- Online booking channels
- Travel agents
- Apps</td>
</tr>
<tr>
<td colspan="3">Cost Structure
- Fixed and cheaper assets
-Labor costs
-Depreciation
- Management costs
- Renovation costs
- 3rd party partner fees
- Investment in technology
- Cleaning services</td>
<td colspan="3">Revenue Streams
- Assets renting (parking for example)
- Management fees
- Customer sales
- Business contracts
- Discounts
- Vouchers
- Stay packages (for X nights one extra night is offered)</td>
</tr>
</table>

5 Conclusions

The present study main goal was to analyse the main impacts of COVID-19 in the hospitality business, more precisely in terms of the necessary business model reinvention always looking at the current and future reality of sustainability. Furthermore, the study also intended to identify the main changes that hotels had to implement to adapt their business in the current pandemic context, basically referring to a reinvention of their business models and to ultimately present a general business model, a post-COVID-19 business model, aiming to contribute to the hospitality industry's reinvention in the global scenario.

To fulfil the research objectives, the study was based on a literature review about the pertaining issue and in-depth interviews conducted with 14 interviewees, who still work at hotels at the present moment. Based on the Business Model Canvas, a business model framework proposed by Osterwalder and Pigneur (2010), the interview consisted of several questions that aimed to assess the nine building blocks that comprise such framework, namely the hotels' key partners, activities, resources, value proposition, customer relationship, channels, customer segments, cost structure, and revenue streams.

The obtained data either supported or contradicted the perspectives presented by Mairinger (2018), Jiang and Wen (2020), Mastrogiacomo (2020), and Sigala (2020), which is why these investigations are the main focus of the present work, especially in the data analysis chapter, where all findings were compared to these sources.

These authors emphasise the negative impacts that COVID-19 has been having on the hotel industry since its emergence, resulting in the loss of profits, decreased reservations, and several staff members being put on lay-off or fired (Sigala, 2020). Jiang and Wen (2020) have suggested that the main three trends at the moment, and in the hotel industry, refer to artificial intelligence, cleanliness, and health and healthcare, essentially summarising the aspects that hotels must consider when reinventing their business model. Mairinger (2018) and Mastrogiacomo (2020), on the other hand, focus on the technological strategies that hotels must implement in their facilities to overcome such crisis and to increase their profits and competitive advantage. Jones and Comfort (2020) mentioned that there are fears that without financial resources the COVID-19 crisis can change or delay the priorities and strategies for sustainability.

It is possible to conclude that the majority of hotels maintained their partnerships with the same tourism agencies, changed their suppliers and cleaning services, and started to develop some activities internally, aiming to ultimately decrease some of the hotels' costs by increasing their demand. Their activities are mainly limited to a small number of clients and some of them are developed on exterior spaces, and whenever possible, due to the imposed social distancing measures. Moreover, hotels had to invest in some other activities, especially those related to the hotel's management and operations, directly associated with new cleaning and safety services, to the training of its staff members regarding the best practices to deal with COVID-19, and to the restructuring of its personnel (some staff members were fired or put on lay-off). The investments made by hotels are significantly related to technology innovations, which demonstrates a clear bet on this dimension to overcome such difficult times, where safety and well-being are the focus.

In addition to these investments in technology, which also aim to attract more customers, hotels also offered personalised services and discounts to customers, to increase their profit. The cancellation policy also became more flexible, with customers being able to cancel reservations up to 24h-72h without further costs or even to reschedule their reservations. Regarding the channels to make such reservations, hotels started to rely on other websites, apps, and partnerships, which suggests that they are now open to new channels to deepen their scope and to attract different customer segments. Their customer segments have significantly changed since they are mainly Portuguese, younger people that wish to stay for several more days than usual. These new strategies were applied to balance their revenue, counterbalancing their costs with their profit.

In sum, this study demonstrates that hotels' future, and specifically after the emergence of COVID-19, truly relies on digital solutions to thrive once again. It has become quite clear that the business model reinvention must be based on technology and digital communication. However, it is noteworthy to mention that the study has some limitations, namely related to the sample size, both in terms of the participants and hotels (14), which might not allow for a wide generalisation of the findings, and as well the difficulty of finding filled a hotel business model. Therefore, further investigations should try to focus on a wider and bigger sample, aiming to better illustrate the current circumstances in more hotels and to fully discuss the COVID-19's impacts on a broader level within the hospitality industry and the necessary reinvention in terms of the hotels' business model. Aspects of human-based services and respective customer requirements should be further investigated in future studies as well as survey various customer profiles. It is also suggested in future studies to specify in more detail the digital strategies. It is also important to mention as a limitation that the study was conducted at a very early stage of the pandemic and without the existence of vaccines, a fact that should be taken into account in future research.

References

Batinic I (2016) Hotel management and quality of hotel services. *Journal of Process Management. New Technologies* 4(1). Centre for Evaluation in Education and Science (CEON/CEES): 25–29

Braun, V., Clarke, V.: Using thematic analysis in psychology. Qual. Res. Psychol. **3**(2), 77–101 (2006)

Chesbrough, H., Rosenbloom, R.S.: The role of the business model in capturing value from innovation: evidence from Xerox Corporation's technology spin-off companies. Ind. Corp. Chang. **11**(3), 529–555 (2002)

Creswell, J.W.: Qualitative Inquiry & Research Design: Choosing Among Five Approaches, 2nd edn. Sage, Thousand Oaks (2007)

Hospitality Technology: Three Ways Technology Will Transform Luxury Concierge in 2018 (2018). https://hospitalitytech.com/three-ways-technology-will-transform-luxury-concierge-2018. Accessed 24 Oct 2020

Ivan, I.: Effects of dynamic organization and digital innovation on the hotel tourism industry during the Coronavirus pandemic period. SSRN Electron. J. (2020)

Jiang, Y., Wen, J.: Effects of COVID-19 on hotel marketing and management: a perspective article. Int. J. Contemp. Hospitality Manage. **32**(8), 2563–2573 (2020)

Johnson, M.W.: Seizing the white space: business model innovation for transformative growth and renewal. Harvard Business School Publishing, Boston (2010)

Jones, P., Comfort, D.: The COVID-19 crisis and sustainability in the hospitality industry. Int. J. Contemp. Hospitality Manage. **32**, 3037–3050 (2020)

Kim, Y.H., Barber, N., Kim, D.K.: Sustainability research in the hotel industry: past, present, and future. J. Hosp. Market. Manag. **28**(5), 576–620 (2019)

Lindgardt, Z., Reeves, M., Stalk, G., et al.: Business Model Innovation When the Game Gets Tough, Change the Game. Boston Consulting Group, Boston (2009)

Lukanova, G., Ilieva, G.: Robots, artificial intelligence, and service automation in hotels. In: Robots, Artificial Intelligence, and Service Automation in Travel, Tourism and Hospitality. Emerald Publishing Limited (2019)

Mairinger, P.: Digital transformation in hospitality: a guidance on how to implement and operate a hotel app to generate incremental revenue and to maximise customer value. Master Thesis, Universidade Católica Portuguesa, Portugal (2018)

Mastrogiacomo, M.: COVID-19 hotel recovery strategy: top 10 digital strategies to thrive in the "new normal" when travel demand strengthens. Industry Update (2020). https://www.hospitalitynet.org/opinion/4098347.html. Accessed 24 Oct 2020

Mitchell, D., Coles, C.: The ultimate competitive advantage of continuing business model innovations. J. Bus. Strategy **24**, 15–21 (2003)
Novelli, M., Gussing Burgess, L., Jones, A., et al.: 'No Ebola…still doomed' – The Ebola-induced tourism crisis. Ann. Tourism Res. **70**, 76–87 (2018)
Okumus, F., Chathoth, P., Altinay, L., et al.: Strategic Management for Hospitality and Tourism, 2nd edn. Routledge, London (2019)
Osterwalder, A.: The business model ontology: a proposition in a design science approach. Ph.D. Thesis, Université de Lausanne, Switzerland (2004)
Osterwalder, A., Pigneur, Y.: Business Model Generation: A Handbook for Visionaries, Game Changers, and Challengers. John Wiley & Sons Inc., New Jersey (2010)
Ritter, T., Pedersen, C.L.: Analysing the impact of the coronavirus crisis on business models. Ind. Mark. Manage. **88**, 214–224 (2020)
Sigala, M.: Tourism and COVID-19: Impacts and implications for advancing and resetting industry and research. J. Bus. Res. **117**, 312–321 (2020)
Sneader, K., Singhal, S.: Beyond coronavirus: the path to the next normal. McKinsey & Company (2020). https://www.mckinsey.com/industries/healthcare-systems-and-services/our-insights/beyond-coronavirus-the-path-to-the-next-normal. Accessed 29 Aug 2020
Teece, D.J.: Business models, business strategy and innovation. Long Range Plan. **43**(2–3), 172–194 (2010)
Tomašević, A.: Luxury hotels: concept and new trends. Bridging Tourism Theory and Practice **9**, 195–211 (2018)
UNWTO: Why tourism? UNWTO (2020). https://www.unwto.org/why-tourism. Accessed 20 October 2020)
Voelpel, S.C., Leibold, M., Tekie, E.B.: The wheel of business model reinvention: how to reshape your business model to leapfrog competitors. J. Chang. Manag. **4**(3), 259–276 (2004)
Wei, W., Torres, E.N., Hua, N.: The power of self-service technologies in creating transcendent service experiences: the paradox of extrinsic attributes. Int. J. Contemp. Hosp. Manag. **29**(6), 1599–1618 (2017)
Yin, R.K.: Qualitative Research from Start to Finish. The Guilford Press, New York (2011)

Software Project Management Myths

Lawrence Peters[1,2](✉)

[1] Universidad Politecnica de Madrid, Madrid, Spain
ljpeters42@gmail.com
[2] Software Consultants International Limited, Auburn, Washington, USA

Abstract. Software Project Management is a topic that is rarely found in the call for papers at Software Engineering conferences. This combined with the fact that most software project managers have not been trained in management methods has resulted in software project managers adopting methods that they believe are beneficial but are not. This is not a sustainable situation as software projects are becoming more expensive with far reaching consequences for failure. This paper cites some of the more common fallacious methods and provides published references refuting their presumed benefits together with suggestions as to how to improve the current state of affairs.

Keywords: Software · Project · Management · Management Fallacies

1 Introduction

The software project manager is an important factor in the success of software projects [1]. In fact, the software project manager has been cited as being more important than all other project success factors combined [2]. That is why the beliefs and actions of the software project manager are so important. For example, if the software project manager believes that a certain set of actions will help the software project team to be successful, they will likely take those actions. If those actions are actually detrimental to the success of the project, it is unlikely that the project will be successful after implementing them. That is why identifying and refuting mistaken beliefs (referred to herein as "myths") is important to the sustainability of software engineering as a profession. A significant barrier to overcome in improving the current situation is the fact that changing one's beliefs is difficult [3].

2 What is a Myth?

There are many definitions for the term, "myth." The one used in this paper that best captures the nature of the thesis here is, "a widely held but false belief." It is false because it cannot be supported by facts and data. The number of myths "widely held" by software project managers will vary from one organization to the next and from company to company. But some subset of these will be present in nearly every organization. The

V. Gupta et al. (Eds.): SSEBIM 2022, LNISO 62, pp. 98–110, 2023.
https://doi.org/10.1007/978-3-031-32436-9_8

collective set of myths held by some software project managers believed to be self-evident truths is probably large enough to be the subject of an entire textbook. However, only the more pervasive, obvious ones will be discussed here. Also, not all software project managers are the same nor do they all subscribe to the same set of myths.

3 Where Did These Myths Come from?

Myths are beliefs that are the result of misinformation, wishful thinking, ignorance, or hearsay and not based on facts and data. Much of this is due to a lack of properly preparing software project managers for their role in software development [4]. The wishful thinking aspect of myths is due to the desire for some simple way to address a complex problem. In the field of software engineering, they can and have resulted in the creation of many antipatterns because we base our actions on our beliefs. An antipattern is an action taken extemporaneously to respond to a specific situation which has occurred in the course of a project. At the time it was taken, the action seemed appropriate, a good idea, but it proved to be detrimental to the project. In order to qualify as an antipattern, this action must have been taken more than once. Nearly 100 of these have been identified and documented [5] with more likely being created all the time. The lack of attention to the needs of software project managers in the various forms of research, publications and studies at the university level has contributed to software project managers having to make up their own approaches (myths and antipatterns) with little or nothing upon which to base the appropriateness of their actions on other than what seemed reasonable at the time [6]. This situation is not totally unexpected since moving from the rank of software engineer to software project manager subjects the new software project manager to situations they may not have experienced or been trained for. At the software engineer level, decisions are based primarily on facts and data. But as one moves to a position of authority, one's intuition influences decision making more and more [7]. The higher in the organization one progresses to, the greater the influence of intuition until intuition is dominant with facts and data being largely ignored [8]. This also explains why software project managers do not seem to learn from project failure. Their intuition guided them and accepting responsibility for a failure reflects on them personally bringing into question their view of their own competence. Instead, the blame is placed on just about any other factor including:

- The requirements changed
- The customer was hard to work with
- Our engineering resources were multiplexed with another project
- The problem was more complicated than we first thought
- The project was underfunded, the schedule too short.
- We were forced to use the Waterfall Lifecycle

Shirking one's responsibility for failure has been shown to disable our ability to learn from failure [9]. Another issue not often suggested in expanded lists of excuses for why software projects fail is if this was the case when the project was being planned, why did you accept a leadership role for it?

4 Some Popular Myths

Management myths are not unique to software project management but exist in various forms in all areas of project management. Due to its relative "youth" and lack of sufficient attention in the literature, software project managers have created and adopted many myths about managing software projects. Some are unique to software development due to its parochial view that software projects are different from all other projects and as such, not subject to the same forces and principles present in other projects. The situation is somewhat complicated by the fact that not all software project managers hold the same set of myths to be self-evident truths. Although new software project management myths may be created with each passing day, the following list enumerates some of the more popular ones including a few that have been published and portrayed as being true. References are provided that refute each myth:

1. *Technology is the key to success in software projects* – A study by IBM found that technology isn't the key to success [10]. After categorizing the documented causes of software project failures, that study found that 53% of the failures were attributable to poor project management while only 3% of the failures could be attributed to technical challenges. This myth is really another form of the "magic bullet" syndrome. That is, the belief that if just the right technology was used, the project would have been successful. Software professionals work with and rely upon technology to accomplish assigned tasks and in many cases for their entertainment. But this dependence on technology tends to spill over into the area of project management with potentially disastrous results. A four year study of 72 multinational product development projects found that project problems that appeared to be technology related actually had, at their source, social, psychological and organizational issues [11]. In retrospect, this seems reasonable since it is people who do the work. If people are not properly managed, organized, coordinated, motivated and so forth, project success is unlikely. To compound matters further for high technology projects, project managers overwhelmingly agree that personnel problems are their most difficult problems to deal with [12, 13] and are the ones they are the least trained for.
2. *People work for money* – For many years, leading experts on human behavior have studied and identified why people work [14, 15, 16]. Although their models differ slightly, one of the things they have in common is that people work for self-esteem, self-realization, and other reasons – not for money. This myth leads software project managers to use money as a means of motivating people to do more work, better work and higher productivity under the presumption that money is what people work for. Money, as an incentive for improving productivity has been shown to work for short periods and only in situations involving repetitive activity such as factory work [17]. It turns out that the best way to improve performance is to thank people for their work [18]. That seems like a paradox in that the most costly reward is less effective than one that is free. The thank you does not have to be some sort of public ceremony. A private, one-on-one meeting between the software engineer and their software project manager is all that is required. Some software project managers do not see the need for saying thank you because (in their words), "I do

not need to thank them for doing the job I am paying them to do." Given what we now know, that position will not motivate people to perform at a high level.

3. *The software project manager must be the best software engineer on the team* – This one has been written about and refuted for nearly half a century [19, 20]. There are three major issues which discredit this one –

    a. The productivity of the team – By putting the most productive software engineer into management, the overall productivity of the team is reduced. This new project manager will be too busy with meetings, reports and personnel matters to spend much time developing software. This results in an overall reduction in productivity.
    b. The mentoring capability of the project manager – Unless this is a very special person, they will not have the patience with poor performing software engineers to assist them in improving their abilities and mentoring them to advance their career.
    c. Effectiveness as a project manager – This high performing person was exceptional not just because they possessed some innate skills but in addition, because they really liked developing software. Taking them away from what they loved doing results in a disgruntled software project manager who would rather be writing code than dealing with the various aspects of software project management. This often results in this skilled individual resigning from the firm, taking a demotion back to software engineer or worst of all staying in this position and making all who report to them as miserable as they are.

4. *Putting pressure on the team will improve the team's performance* – This one seems reasonable, implying there will be negative consequences to the team if the team is not successful. The theory is that this pressure should get everyone to make an extra effort. We frequently see this strategy in books and movies. In fact, people work at lower productivity levels when they are under stress [21]. Also, if there is enough pressure placed on the team, the collective knowledge and experience of the team will suffer. Under such circumstances, team members work independently, not relying on other team members to help them with a particularly troublesome software problem or helping other team members with theirs [21]. The result? The collective knowledge of the team is lost, team members work independently of each other and the productivity of the team overall suffers [21]. How much pressure is too much pressure? That is one of the "soft" issues successful software project managers have mastered. By maintaining open communications with the team, an effective software project manager can tell when the team is not performing at its highest level and work to reduce the pressure in some way.
5. *The best team is composed of the best software engineers available* – This concept has been tried in other high technology endeavors and it has consistently failed [22]. This phenomenon even has a name, "The Apollo Syndrome." In part, what happens is we have a team of (for lack of a better term) "prima donnas" some believing they are smarter and more accomplished than the others. Under these circumstances, they do not work together as a mutually supportive team. Instead, they break up into individual contributors and/or separate factions often competing

with each other in order to demonstrate their superiority. The result is detrimental to the project.

6. *Offer a big reward to get higher productivity* – People do not work for money but if the reward is big enough, they will cheat in order to get the reward [23]. Big rewards (e.g. a trip to some place special, paid for by the company) can undermine ethical behavior by one or more team members resulting in friction within the team and loss of productivity.
7. *Start with a bigger team* – Brooks' admonition to not add people to a late project has caused some software project managers to conclude they should start with a larger team than they and their senior technical adviser(s) believe they really need. They believe this strategy (though more costly) will prevent them from getting behind schedule. Empirical and field studies [24] have shown that larger teams are less effective than "right sized" teams. Larger teams require more coordination, more supervision and more interactions within the team. Overall, larger teams have been shown to be less efficient and productive.
8. *Spread success by distributing members of successful teams to other teams* – Empirical and field studies [25] have shown that breaking up a team to do this results in the loss of an effective team and the intended spreading of the successful concepts does not happen. If the goal really is to spread successful practices, then study what the manager of that project did and emulate it elsewhere. More than this, a study of 1,004 development projects involving more than 11,000 people found that when team familiarity (i.e. team members had worked together before) increased by 50%, defects decreased by 19% and budget deviations decreased by 30% [26].
9. *Planning and scheduling are the same* – Contrary to some published opinions giving this one the imprimatur of apparent truth [27], this is simply not the case. It can be easily refuted. Planning and scheduling are related to be sure but not the same. A plan is simply a list of tasks and subtasks that must be successfully completed for the project to be deemed a success. A schedule time orders the tasks indicating which must be done first, second and so on as well as which can be done in parallel. These two aspects of the project (the plan and the schedule) are, in a sense, linked by the assignment of specific individuals to tasks. Since planning is a continual activity throughout the life of the project [28], the schedule may also need to be revised frequently as well.
10. *Coming up with the right plan/lifecycle helps ensure success* – This is another wishful thinking or "magic bullet" concept. The problem is that successful project managers realize and act upon the notion that the initial plan is only the start. In fact, planning is a continuous activity throughout the life of the project [28]. As General Dwight Eisenhower was quoted as saying, "Plans are nothing, planning is everything" indicating that planning is a continuous activity. The reason why it is continuous is that, throughout the project, unforeseen problems will occur that require a change in the project plan and schedule as some tasks occur late while others go unexpectedly smoothly, taking less time and effort than anticipated but potentially causing a coordination problem. As the Greek philosopher Heraclitus (535 BC – 475 BC) was quoted as saying, "Change is the only constant."
11. *Requirements changes cause software project failures* – Requirements changes are almost inevitable in nearly all software and other technology related projects

because requirements definition is a "discovery" process. The project begins with all the stakeholders possessing some vision of what will be produced. That vision skews their understanding of what the requirements are. Over time, it becomes obvious to some that what they thought was going to be delivered differs significantly from what is likely to be delivered. This results in requirements changes. In the construction industry, requirements changes are a way of life [29, 30]. The changes themselves in any industry are usually not the problem. Not planning for and accommodating changes in a cost effective manner is. Assuming there will be no changes in requirements and other simplifying assumptions put the project at risk. Changes can cause rework increasing cost and schedule which can put the project at risk if not accommodated in the project plan.

12. *If our project goes over budget or gets behind schedule early on, we can work harder and eventually finish on budget and on schedule* – There are, literally, no facts and data to support this one. What facts and data that do exist, based on a study of over 700 projects indicate that if the project is 15% complete and over budget and/or behind schedule, its chances of recovering and finishing within its budget/schedule are nil [31]. This emphasizes the need for the software project manager to closely monitor cost and schedule right from the start of the project and being willing to take remedial action if the project begins to depart from the plan and schedule.
13. *Quality results will cost more* – This one seems odd since we have known for decades that having to correct software errors after the system is delivered or goes live is much more expensive than reducing the error count before delivery [1]. But papers keep being published that suggest that quality is more expensive than the lack of it [32]. The desire to produce quality results drives people to higher levels of productivity [33, 34] because everyone wants to produce quality results and wants to be associated with a quality product. When people are prevented from doing so based on a practice of just shipping something regardless of its quality because the schedule calls for it, they are in a state of cognitive dissonance and produce at their lowest level of productivity [11, 35].
14. *Software engineering is unique in that budgets and schedules are rarely met* – The same is true for many ground breaking construction projects as well. This is particularly true when attempting to build something that has never been attempted before. However, there have been some notable exceptions. For example, in the 1930s, Hoover Dam was the largest structure of its kind ever attempted. In spite of stringent requirements, a challenging completion date and terrible working conditions (e.g. desert heat, the needed machinery did not exist and had to be invented/designed/developed) it finished under budget and ahead of schedule. *Post partem* analysis of this project revealed that the five company consortium responsible for it engaged in extensive planning and scheduling activity long before and during project execution. Even though man has been building roads and bridges that are seemingly unchallenging tasks for more than two thousand years, overruns in budget and schedule occur today with a great deal of regularity [36]. Given all that we should have learned over the centuries, this seems puzzling until we become cognizant of the Nobel Prize winning work of Daniel Kahneman who won the Nobel

Prize in Economics for explaining this phenomenon of inaccurately estimating as a common trait shared by all human beings [37].

15. *If we had better estimating methods, we would come closer to meeting budget requirements* – It turns out that no matter how hard we try to accurately estimate any project, we are unknowingly placed at a disadvantage. The problem lies, not in our methods but in our selves. Research into how people estimate [37] found that we are overly optimistic about our abilities while focusing on the benefits of the project's result so much that we ignore or downplay the risks involved. This human trait is present no matter what formulated method we use [38]. More recently, a method has become available that helps us back out the effects of over optimism and bound our estimate [29, 36]. It works so well that the American Planning Association has advised its members to never use traditional estimating methods without also using this method called, "Reference Class Forecasting." What it provides is an estimate plus a set aside or contingency amount which is based on the desired confidence level for the estimate [29, 36].
16. *Not using the Waterfall Lifecycle improves a project's chances of success* – Anecdotally, this lifecycle model has been taken to task for not being viable in today's software engineering environment. At the same time, an alternative lifecycle with facts and data to support the contention that this new lifecycle model enhances a project's chances of success has not been promoted. Agile and other approaches have been anecdotally described as being an improvement without overwhelming evidence. What is needed here are a set of facts and data. Until a study is performed looking at hundreds or thousands of software projects, categorizing their lifecycle models and clearly demonstrating this one is actually true, it must remain a myth. Besides, what we do have strong evidence for is placing the success or failure of a software project squarely on the software project manager, not the lifecycle employed [10].
17. *We can multiplex our best people* – With the increase in work and complexity in so many organizations, it has become common to share highly skilled workers with more than one project. The simplest sharing arrangement being 50–50. That is, having an individual spend half of their working hours on one project and the other half on another project. The presumption is that since people are highly adaptive, between the two projects, the organization will still get 100%. That may sound reasonable to some but those who have been the worker making such multiplex commitments will attest to the fact that it simply does not work that way. People are not "plug compatible" and able to instantly switch from one set of project issues to another without some form of "spin up" taking place. That is, one has to catch up as to where things are now versus where they were the last time they worked on this effort. Thus, the organization is not going to get a total of 100% from the software engineer. How much they will get will vary depending on several factors but it will not be 100%.
18. *Set challenging goals, if people have a target, they will work to achieve it* – As with some of the other myths listed here, this one is a question of degree, not an absolute. Senior managers often like to call the goals that are set for an organization, "stretch goals." This name implies what the senior manager wants to achieve. That is, to get the team to push themselves to increase or "stretch" their performance. As well

intended as this concept may be, studies have shown that if the team perceives the goal as being unachievable, productivity suffers [39]. So the admonition here is to set goals that are achievable. Finding out what the team thinks is achievable must be part of the software project manager's communication skill. Also, setting a goal and achieving it increases the brain's oxytocin resulting in higher productivity [40].

19. *If it isn't broken, don't fix it* – This one is popular in many industries but, even though it sounds reasonable, it sets up a culture of maintaining the *status quo*. A better approach is to proactively establish a culture of continual improvement [41]. A slight revision of this philosophy to a proactive one results in the revised statement, "If it isn't breaking, don't fix it." This imparts a philosophy of continually monitoring and improving processes, methods, techniques and related matters to do better as a team going forward. If the change does not improve matters, undo it and try something else. Even highly successful athletes continually seek to improve their performance. The concept here is to constantly monitor what is happening in our project(s) looking for ways to fine tune our methods to reduce cost, improve quality and increase our proficiency. Based on studies of how software engineers view their work and relationship to management [13], such an environment motivates software engineers to higher levels of productivity.

5 Are Software Project Managers Really Needed?

In the early years of its corporate life, the founders of Google™ subscribed to the thesis that they really did not need managers. They created a "flat" organization in order to prevent obstacles to innovation and creativity. They were hoping to create the kind of collegial environment their software engineers experienced in their university years. In a few months, they abandoned this approach [42]. Details of how to address one issue or another all ended up being sent to the founders of the company, overwhelming their ability to maintain any form of control. Over time, Google's founders came to conclude that managers did more than sign expense reports and answer questions but communicated corporate strategy to the thousands of software engineers employed by the company and contributed in other ways as well. These contributions included prioritizing work, helping software engineers to develop their careers, motivating individuals and teams and keeping everyone focused on corporate goals [42]. So, we really do need software project managers after all.

6 Removing Myths from the Managers' Lexicon

The easy solution for dispatching these and other myths about software project management lies in three words – attention, information and education. The more difficult part of the solution is the natural follow-up question – how do we accomplish this? This paper contains several references which can serve as resources to aid the instruction question. But most of the resources outside of this list (e.g. most books on software project management) demonstrate that we have a great deal to learn regarding what software project management involves. In a word, software project management involves "soft" topics. These include organizational behavior, psychology, risk management, complexity management, accounting basics, planning, scheduling, estimating and others which have been described as being "wicked" problems [29, 43] in that they do not have right or wrong answers and seem to change continuously. But consistent with the concept of wicked problems, these myths are merely symptoms of a much higher level problem. That problem is our failure to elevate the status and importance of software project management to being a vital part of the software engineering profession. One example, is the fact that many texts on software project management are mostly about programming, thus reinforcing the notion that software engineering technology is the key to a successful software project. As demonstrated earlier in this paper, it isn't. Another example is the latest version of the Guide to the Software Engineering Body of Knowledge (SWEBOK) [44]. It devotes approximately 11 pages to each of the various aspects of software development and about the same amount to software project management. This seems a bit out of balance given that software project management is multifaceted and has been described as being, "more important to the success of software projects than all other factors combined" [2]. Given its relative importance and the distinct differences between the nature of software development and software project management, more space or, better yet, a separate body of knowledge should be devoted to software project management rather than an extension of the Project Management Institute's PMBOK (The Guide to the Project Management Body of Knowledge) [45]. In addition to the issue of textbooks, and referential resources, the experience of instructors can introduce some credibility issues unless they have industrial experience managing software projects of various sizes.

7 The Response of Software Engineers

After experiencing the seemingly mindless actions of software project managers, some software engineers have concluded that software project managers are really not up to the task. The problem is that most software engineers do not know what a software project manager's task is. It is not primarily technical in nature. This leads to misjudging the behavior of the software project manager. Communicating, coaching, encouraging the members of the team constitute the software project manager's primary activities. A software project manager's failure to concur with a software engineer's technical analysis is a clear rejection of facts and data [7] which, in the eyes of software engineers relegates that software project manager to the category of incompetent. Although educating current and future software project managers is important, their effectiveness in managing projects can be undermined if software engineers are not also educated regarding the important role the software project manager plays in software projects. That knowledge can only serve to improve communications between the software project manager and the rest of the team with incumbent improvement in project success. It may also help to prevent some software engineers from becoming software project managers for the wrong reasons (e.g. more money, prestige, better perquisites). The ones that do will at least know in advance just what they are getting into.

8 What Makes a "Good" Software Project Manager?

We have looked at the mistaken beliefs held by many software project managers and what can be done to eliminate those beliefs but does removing them from the belief systems of software project managers, make them "good" managers? What a good manager is has been known for at least half a century since Peter Drucker [46] first began developing the modern concept of management. Some exemplary work done at Google regarding management practices has produced the best results. They have catalogued these practices [42] for us. A good manager is someone who:

1. Is a good coach
2. Empowers the team and does not micromanage
3. Expresses interest in and concern for team members' success and personal well-being
4. Is productive and results oriented
5. Is a good communicator – listens and shares information
6. Helps team members with career development
7. Has a clear vision and strategy for the team
8. Has key technical skills that help him or her to advise the team

Note that technology issues are present in only one of the eight skill areas listed above further emphasizing the notion that software project management is not primarily a technical activity. While the preceding could all be elaborated on, they make the case for soft skills training which has caused at least one author to observe that, "… Managing remains under studied and under taught" [42]. There are many reasons for this situation including:

- Not recognizing the value that competent software project management provides to the project and organization overall – For example, turnover (i.e. people leaving the project and/or company) is very expensive. As much as 60% of project cost [47] can be attributed to turnover. The number one reason for people leaving the project or company is – management! Hence, one way to reduce project cost is to reduce turnover by improving the quality of software project managers including their ability to help software engineers to advance in their careers.
- Not knowing precisely what it is software project managers do – The word, "precisely" is used intentionally here since management activity runs the gamut from listening to complaints to reviewing the performance of software engineers, helping them advance their careers and more. The fact that most companies, even major corporations, do not have a clearly defined path to management highlights this lack of understanding of this issue [12].

Lack of experience in industry by those tasked with teaching software project management. Software project management is not just "messy" (i.e. poorly structured, variable from one project and company to the next) it is challenging to the software project manager in ways they have not been trained for with little insight into whether the action(s) they take are correct until it is too late. A further complication is that it deals in a domain composed of "soft" issues (e.g. motivation, personnel selection, team management, performance evaluation). The adage, "We teach what we know" applies here. Assistance from accomplished software project managers from industry could help fill this need.

9 Closing Comments

For more than half a century, software engineering has pursued a course of action focused on technology development to solve software engineering problems. Thus far, the technology domain has produced some improvements in quality and a linear increase in productivity of approximately one source line per programmer month per year from 1960 to 1990 [48]. Presumably, that linear increase has continued. Today, some software projects continue to fail. They finish late, are over budget, do not meet requirements or are cancelled. It appears that technical solutions are not where sustainable solutions lie for software project problems but we continue to pursue them, unable to shake our early belief in technology. Changing one's beliefs is very difficult [3]. It has been shown that the management of software projects is where we can obtain the highest leverage (18 to 1) [9] and the greatest return on investment if only we turn our attention to it. A start might be to include software project management as a required course in undergraduate and graduate software engineering degree programs and include it as a topic in the call for papers in every software engineering related conference. Training in software project management must be focused on the people related issues (e.g. team formation, motivation, inter-personal interactions, organizational behavior) as well as the business related ones. Currently, software project management is not a required course in most under graduate and graduate curricula nor is project management a topic area in the call for papers at software engineering conferences, including this one.

References

1. Boehm, B.: Software Engineering Economics, pp. 486–487. Prentice-Hall, Englewood Cliffs (1981)
2. Weinberg, G.: Quality Software Management, vol. In: New York, N.Y. (ed.) 3: Congruent Action, pp. 15–16. Dorset House Publishing (1994)
3. Festinger, L., Riecken, H., Schachter, S.: When Prophecy Fails. Martino Publishing, Mansfield Centre, CT (2011)
4. Tomer, A.: Software management teaching project from software engineer perspective. In: Global Engineering Education Conference (EDUCON) (2014)
5. Laplante, P.A., Neill, C.J.: Antipatterns: Identification, Refactoring, and Management. Taylor & Francis, Boca Raton (2005)
6. Silva, P., Moreno, A., Peters, L.: Software project management: learning from our mistakes. IEEE Softw. **32**(3), 40–43 (2015)
7. Taylor, B.: Why do smart people do such dumb things? Harvard Bus. Rev., 11 January 2011
8. McAfee, A.: The future of decision making. Harvard Business Review, 07 January 2010
9. Myers, C., Staats, B. and Gino, F.: My bad! how internal attribution and ambiguity of responsibility affect learning from failure. Working Paper, pp. 14–104. Harvard Business School (2014)

10. Gulla, J.: Seven reasons IT projects fail. IBM Syst. Mag., February 2012
11. Thamhain, H.: Changing dynamics of team leadership in global project environments. Am. J. Ind. Bus. Manag. **3**(2013), 146–156 (2013)
12. Maylor, H., Turner, N., Murray-Webster, R.: How hard can it be?. Research-Technology Management, pp. 46–51 (2013)
13. Katz, R.: Motivating technical professionals today. IEEE Eng. Manage. Rev. **41**(1), 28–37 (2013)
14. Herzberg, F.: 1966, Work and the Nature of Man. The World Publishing Company, Cleveland (1966)
15. Maslow, A.H.: The Farther Reaches of Human Nature. Viking Press, New York (1971)
16. McClelland, D.C.: The Achieving Society. Van Nostrand-Rheinhold, Princeton (1961)
17. Ryan, R.M., Deci, E.L.: Intrinsic and extrinsic motivations: classic definitions and new directions. Contemp. Educ. Psychol. **25**(2000), 54–67 (2000)
18. Grant, A.M., Gino, F.: A little thanks goes a long way: explaining why gratitude expressions motivate prosocial behavior. J. Pers. Soc. Psychol. **98**(6), 946–955 (2010)
19. Townsend, R.: Up the Organization – How to Stop the Corporation from Stifling People and Strangling Profits. Alfred Knopf, New York (1970)
20. Townsend, R.: Further Up the Organization – How to Stop Management from Stifling People and Strangling Profits. Alfred Knopf, New York (1984)
21. Gardner, H.K.: Performance pressure as a double edged sword: enhancing team motivation while undermining the use of team knowledge. Working Paper, pp. 09–126. Harvard Business School, January 2012
22. Belbin, R.: Management Teams – Why They Succeed or Fail. Butterworth Heineman, London (1996)
23. Gino, F., Ariely, D.: The dark side of creativity: original thinkers can be more dishonest. J. Pers. Soc. Psychol. **102**(3), 445–459 (2011)
24. Staats, B.R., Milkman, K.L., Fox, C.R.: The team sizing fallacy: underestimating the declining efficiency of larger teams, Forthcoming Article in Organizational Behavior and Human Decision Processes
25. Staats, B.R., Gino, F., Pisano, G.P.: Varied experience, team familiarity, and learning: the mediating role of psychological safety. Working Paper, pp. 10–016. Harvard Business School (2010)
26. Huckman, R., Staats, B.: The Hidden Benefits of Keeping Teams Intact. Harvard Bus. Rev. **91**, 27–29 (2013)
27. McConnell, S.: The software manager's toolkit. IEEE Softw. **17**, 5 (2013)
28. Peters, L,. Moreno, A.: Educating software project managers – revisited. In: Conference on Software Engineering Education and Training, Florence, Italy (2014)
29. Peters, L.: Managing Software Projects on the Edge of Chaos – From Antipatterns to Success. Software Consultants International Ltd., Auburn (2015). Kindle eBook
30. Touran, A.: Calculation of contingency in construction projects. IEEE Trans. Eng. Manage. **50**(2), 135–140 (2003)
31. Fleming, Q., Koppelman, J.: Earned Value Project Management –, 4th edn. Project Management Institute, Newtown Square (2010)
32. Petre, M., Damian, D.: Methodology and culture: drivers of mediocrity in software engineering? In: Foundations of Software Engineering Conference, Hong Kong, China, vol. 14, pp. 829–832, 16–21 November 2014
33. Crosby, P.: Quality is Free. Signet, New York (1980)
34. Crosby, P.: Quality is Still Free. McGraw-Hill, New York (1995)
35. Weinberg, G.:, The Psychology of Computer Programming – Silver Anniversary Edition. Dorset House, New York (1998). (originally published in 1971)

36. Flyvberg, B.: From nobel prize to project management getting risks right. Proj. Manage. J. **37**, 5–15 (2006)
37. Kahneman, D.: Nobel Prize in Economics (2002)
38. Lovallo, D., Kahneman, D.: Delusions of success: how optimism undermines executives' decisions. Harvard Bus. Rev. **81**, 56–63 (2003)
39. Ordonez, L., Schweitzer, M., Galinsky, A., Bazerman, M.: Goals gone wild: the systematic side effects of overprescribing goal setting. Acad. Manage. Perspect. **23**(1), 6–16 (2009)
40. Zak, P.J.: "The Neuroscience of Trust," from Management behaviors that foster em, ployee engagement, Harvard Bus. Rev., January–February 2017
41. Weick, K.: Organizational Culture as a Source of High Reliability. Calif. Manage. Rev. **29/2**, 112–127 (1987)
42. Garvin, D.: How Google sold its engineers on management. Harvard Bus. Rev. **91**, 74–82 (2013)
43. Peters, L.: Getting Results from Software Development Teams. Best Practices Series, Microsoft Press, Redmond (2008)
44. IEEE Computer Society: Guide to the Software Engineering Body of Knowledge (SWEBOK), Version 3.0 (2014)
45. Project Management Institute: Guide to the project management body of knowledge. In: Proceedings of the 5th Edition, Project Management Institute, Newtown Square, PA (2013)
46. Drucker, P.: Managing for Results (Reissue). Harper Business, New York (2006)
47. Cone, E.: Managing that churning sensation. Inf. Week **680**, 50–67 (1998)
48. Jensen, R.: Don't forget about good management. CrossTalk Mag., 30, August 2000

Interoperability Between Health Information Systems

João Fernandes[1], Carlos Jeronimo[1], Leandro Pereira[2], Alvaro Dias[2], Renato Lopes da Costa[2(✉)], and Rui Gonçalves[1]

[1] ISCTE-IUL, Av das Forças Armadas, Lisbon, Portugal
{joao_gabriel_fernandes,carlos.jeronimo}@iscte-iul.pt
[2] BRU-Business Research Unit, ISCTE-IUL, Av das Forças Armadas, Lisbon, Portugal
{leandro.pereira,alvaro.dias,renato.Lopes.Costa}@iscte-iul.pt

Abstract. In the health system, the adoption of Informatic and Communications Technologies (ICTs) are becoming visible, with more heterogeneity and interoperability between Information Systems (IS) and medical data. The technological revolution enhances interoperability between Health Information Systems (HIS) as a new challenge to achieve sustainability. Thus, this study aims to explore and analyse the communication between information systems in Portuguese hospitals.

This study follows qualitative research, sixteen semi-structured interviews were conducted with healthcare professionals and hospital suppliers to collect their experience on interacting with HIS, the stakeholders' role, and the measures that are considered crucial to improve the current paradigm. The subsequent analysis was supported with the MAXQDA program to organize and identify the interviewees' keywords, explanations, and views.

The key findings of the study were the inadaptation of public hospitals' board to manage HIS and promote sustainability, the importance that healthcare professionals have in the implementation of new HIS in hospitals and the lack of interoperability between the systems resulting in slower and complex navigation for the user.

To achieve successful interoperability between HIS, hospitals need to have a clear hospital strategy focused on the digitalization of processes. Providing an IT structure that supports the implementation and management of the systems, managing the relationship with the hospital's suppliers, and encouraging the healthcare professionals to involve and use this technology. Further research needs to consider a larger sample and other research methods to identify new constraints and opportunities about this topic.

Keywords: Interoperability · Health Information Systems · Portuguese National Health System · Sustainability

1 Introduction

Digital health can support the healthcare sector to achieve the ambitious health goals of Sustainable Development Goals – SDGs (Asi,Y. & Williams, C., 2018). The World

V. Gupta et al. (Eds.): SSEBIM 2022, LNISO 62, pp. 111–121, 2023.
https://doi.org/10.1007/978-3-031-32436-9_9

Health Organisation created the term eHealth that involves the activities that use ICTs for health services which include the Clinical Information Systems (CIS) that encompasses the electronic medical records, decision support, and monitoring of health systems practices (World Health Organisation, 2016). To evaluate the technological capabilities installed in the healthcare centres, it was developed the Electronic Medical Record Adoption Model (EMRAM) by HIMSS divides the adoption and utilization of EMRs functions into eight stages that start with "Stage 0" until "Stage 7". The lowest stage considers that the organisations do not have any Electronic Medical Record (EMR) system installed in the department systems (laboratory, pharmacy, and radiology) and the highest stage represents the institutions that consolidated the EMR adoption.

The successful implementation of Health Information Systems (HIS) has an impact on the safety of medical treatments and quality of healthcare by increasing the productivity in hospitals that establish more efficient procedures and processes and improvements in the quality of patient's data collection and management (Kruse et al., 2018) (Stavert-Dobson & Risk, 2018). This technology enables a more simplified communication between healthcare professionals with less paper flow, easier access to medical knowledge, information sharing, and monitoring the patient's vital data plus the diminution of human errors (Dobrzykowski & Tarafdar, 2017) (Reza et al., 2020). In the post-implementation process, there is a loss of productivity due to the necessary adaptation of users to operate with the system and adapt the healthcare processes which represents time-consuming and regular support to healthcare providers. (Kruse et al., 2018) (Martins et al., 2019).

Even though the HIS demonstrate to have a positive impact on hospital context, there are visible obstacles due to the lack of organizational resources and high investments necessary to operate with these systems aligned with the resistance of healthcare professionals during the change process (Adler-Milstein et al., 2017) (Marto, 2017). Therefore, is crucial to establish a healthcare professional-hospital administration relationship based on shared value and provide the educational and other supportive resources to stimulate the integration of HIS among the operational and administrative staff (Dobrzykowski & Tarafdar, 2017) (Stavert-Dobson & Risk, 2018).

The heterogeneity of HIS operating in hospitals aligned with the absence of interoperability standards to establish a communication flow between EHRs, patients, and hospitals is creating a barrier to effective Information Technology (IT) intervention in healthcare treatments (Samal et al., 2016) (Kruse et al., 2018). Even though the HIT provides a structure that allows consistent data capture, the issue persists since the data tends to be rarely standardized, fragmented, and stored in multiple platforms with incompatible structures (LeSueur, 2017) (Aceto et al., 2018). Therefore, there is a decrease in the effectiveness and decision-making of healthcare professionals due to the higher complexity in accessing and analysing the information stored in diverse databases (Dobrzykowski & Tarafdar, 2015) (Jardim & Martins, 2016).

The interoperability between HIS allows the exchange of information between systems from different suppliers and creates an open-source of shared services in healthcare organisations (Dodd, 2017) (Frisse, 2017). Organisations like Health Level Seven International (HL7) are responsible to design standards that allow healthcare information transmission between IS which can be made via point-to-point or via middleware

(Oemig & Snelick, 2016) (Kuo & Kuo, 2017). Interoperability enables faster, reliable, and consistent clinical data sharing (Longo, Dan L.; Drazen, 2016) which improves patient safety and cost reduction in treatments and stimulates medical research (Oemig & Snelick, 2016) (Peterson et al., 2016).

In the Portuguese context, the Ministry of Health (MoH) owns and operates in the majority of hospitals and community-based clinics to provide equal access to all citizens to healthcare services (de Almeida Simoes et al., 2017). The MoH centralizes the responsibility for the most relevant process of healthcare services which enhances the political dependence of the national health system. Regarding the implementation and development of HIS is visible the lack of private players operating in this market, difficulties in innovation processes, and the absence of a clear analysis of the IS operating in the public healthcare units (Sousa, 2017). The public hospitals have the opportunity to operate with free of charge HIS developed by SPMS (Serviços Partilhos do Ministério da Saúde). The most used systems are SONHO more related to administrative functions and SClínico a single and common EHR to all healthcare providers that allow access, use, and share of clinical records[2]. About the communication between systems, the SONHO V1 establishes a Database Links (DBLINK) connection with the HIS from other suppliers which is a more complex process since it is needed to develop individual schemes of integration for each hospital (Marto, 2017). Even though the hospitals have systems to record information with common aspects some problems appear in effectively connecting the hospital's data with institutional databases, the absence of an integration mechanism for non-Government HIS, and lack of a common unique identifier for patients (Jardim & Martins, 2016).

Regarding the lack of interoperability between HIS, it is considered an obstacle to creating an efficient flow of information about patient's clinical (Fan et al., 2018), a loss of time in the clerical processes of healthcare professionals (Samal et al., 2016) and can lead to clinical errors, such as medication and diagnostics lapses (Poly et al., 2018). The poor user interface design between systems and not the intuitive layout of EHRs may lead to healthcare professionals' errors in data input or information comprehensions (Sittig et al., 2020) aligned with the possible lack of ICTs skills of system's users (Aceto et al., 2018). These issues are linked with technical debt design decisions leading to coupling and cohesion issues which impacts the sustainability of software architectures (Colin C. Venters et al., 2018).

2 Materials and Methods

Qualitative research allowed this study to explore the current level of interoperability in Portuguese hospitals, identify the key aspects to establish interconnection between HIS, and understand the influence of interoperability or absence has on the healthcare professional's routine in Portugal. The main goal was to analyse the importance that interoperability between IS generates in healthcare.

For this purpose, semi-structured interviews allowed to retract the use of IS during hospital procedures, collect the feedback from healthcare professionals that use IS, information about the hospital management, the relationship with the stakeholders, and the measures that consider essential to improve the current paradigm. The persons interviewed were selected based on their functions in the healthcare sector, their experience, and their expertise in HIS. In terms of geography, the sample is limited to Portugal and mainly interviews from public hospitals so that should be avoided generalizations. Sixteen interviews were conducted with seven healthcare professionals from four different hospitals (three physicians, three nurses, and one radiologist) and nine members of four hospital's IT and process suppliers (Colin C. Venters et al., 2018).

About the organisations involved in the research, Company I is a consulting company that has been involved in projects to support the implementation and modifications of HIS. On the other hand, Company II is a public entity responsible to supply IS on the majority of NHS hospitals, Company III is the private IS supplier with a high market share in Portuguese hospitals, and lastly, Company IV is a smaller company that focuses on developing smaller EMRs. The eight hospitals mentioned have a different number of beds available for healthcare treatments and different IS suppliers. Hospitals I, V, VI, VII, and VIII are public institutions that operate with the IS from Company II, and Hospital II is also public, but their main supplier is Company III. On the other hand, Hospitals III and IV are part of the most advanced Portuguese hospitals in terms of technology, these private organisations operate specially with IS from Company III.

3 Data Analysis and Results

The interview records were transcript in separate documents introduced in the MAXQDA program that allowed to track the keywords, expressions, and descriptions mentioned during the interviews and create a level of category based on the specific details mentioned in the selected items from the interviews. The selection of explanations that characterized the HIS operating in the healthcare units and their interaction during their daily routines allowed to contextualize the hospital's context.

3.1 Impact of IS on Hospital Context

Each interviewee described their experience with HIS and the impact that it brought to their routines. The main conclusion was that HIS promoted the dematerialization of processes in hospitals which is known as process digitization. Furthermore, it was identified that 94% of the participants characterized the use of HIS as complex and a cause of lost time in their routine and 64% enhanced that hospital structure is not prepared for the technological transformation (see Fig. 1).

3.2 Dematerialization of Processes

The consequences generated by the dematerialization of processes (see Fig. 2), the main advantages pointed out by the interviewees were the increase of information access and an efficient analysis from the data storage and was reported the active support given to the healthcare professionals when interacting with HIS. Even though it was not considered the most significant consequence of digital processes, the reduction of paper flow was pointed out by 44% of the participants as a positive change in their daily routine.

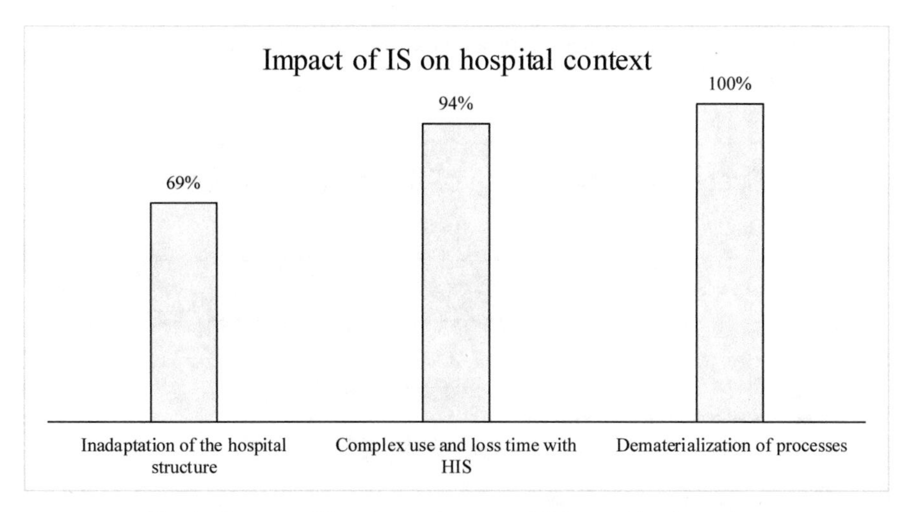

Fig. 1. Impact of IS on hospital context (Elaborated by the author)

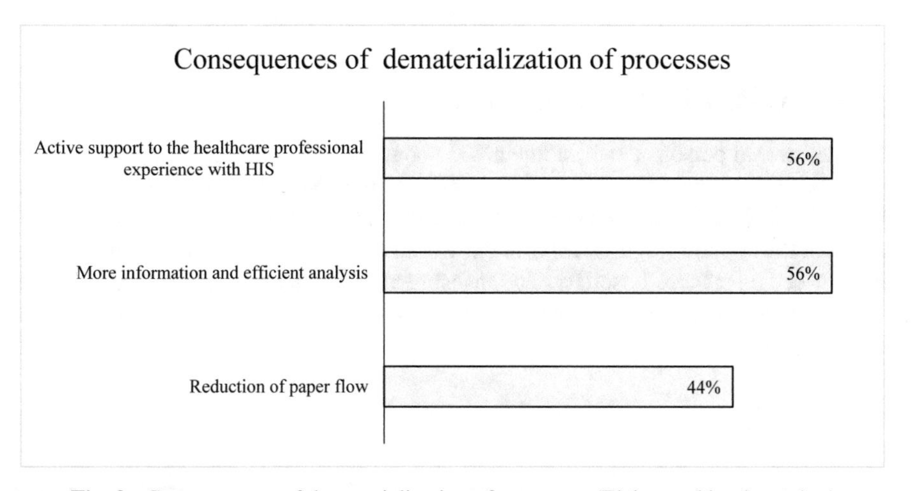

Fig. 2. Consequences of dematerialization of processes (Elaborated by the author)

3.3 Inadequacy of the Hospital Structure

Figure 3 shows that most of the people interviewed described difficulties from healthcare providers in adapting to the digital transformation. One of the supplier's members referred that "healthcare professionals tend to be resistant to change, so it is needed to drag them among the changing process" which demonstrates the need to involve them in the HIS implementation process. Also, 45% mentioned the low-efficient performance from the public hospitals in dealing with the digital processes focusing on the weak management from the administration and the inefficient IT department role with suppliers. An interviewee pointed out that in public hospitals "there is no hospital strategy, they have a vast number of IS suppliers and constantly purchasing new systems just for specific functionalities".

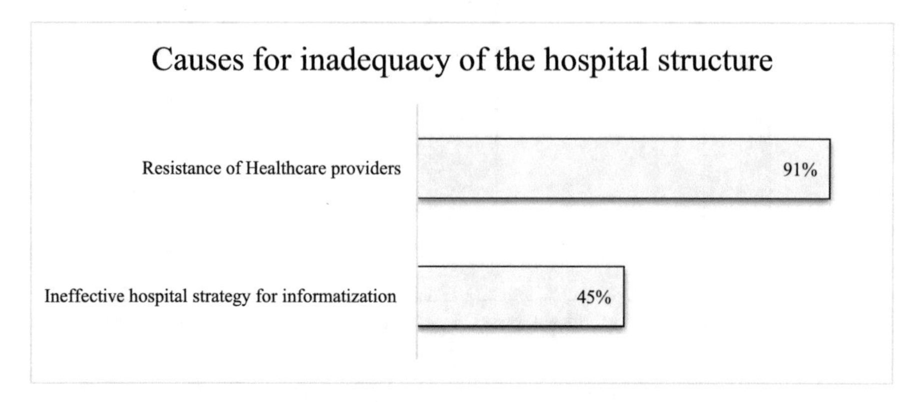

Fig. 3. Causes for inadequacy of the hospital structure (Elaborated by the author)

3.4 Complex Use and Loss Time with HIS

The interviewed persons (91%) complained about the lack of integration between HIS among healthcare providers, IS suppliers, and external consulting managers. Also, 64% of them refer that HIS layout is not user-friendly and not intuitive and insufficient support from the suppliers and IT department from the hospital. Even though just 24% of interviewees refer those hospitals need investment in IT services, taking into consideration that their testimony enhanced as a problem faced on the hospital from the public sector (see Fig. 4).

4 Discussion

Regarding the strategy for HIS in Portuguese hospitals, 69% of the interviewed people mentioned the inadequacy of the hospital structure to deal with IS, and 45% consider that the Portuguese hospitals have an ineffective strategy. According to the interviews, the public hospitals have a low-efficient strategy due to weak administration management, inefficient IT department performance, and the lack of investment in computer structures. The hospital structure was characterized by having IT departments that are not prepared to play the intermediate role with the suppliers and a lack of workgroups for the hospital's digitalization. The Hospital III (Stage 7 of the EMRAM scale) and Hospital IV are examples of a successful strategy on HIS that brought benefits, such as the reduction of patients length of stay (van Poelgeest et al., 2017), an improvement on the quality of care management and treatments (Stavert-Dobson & Risk, 2018). Both entities have a private investment with a high level of integration between the systems supported by Company III. So, was possible to understand that there is a clear strategy from the private hospitals, but what about the issues described in the public hospitals?

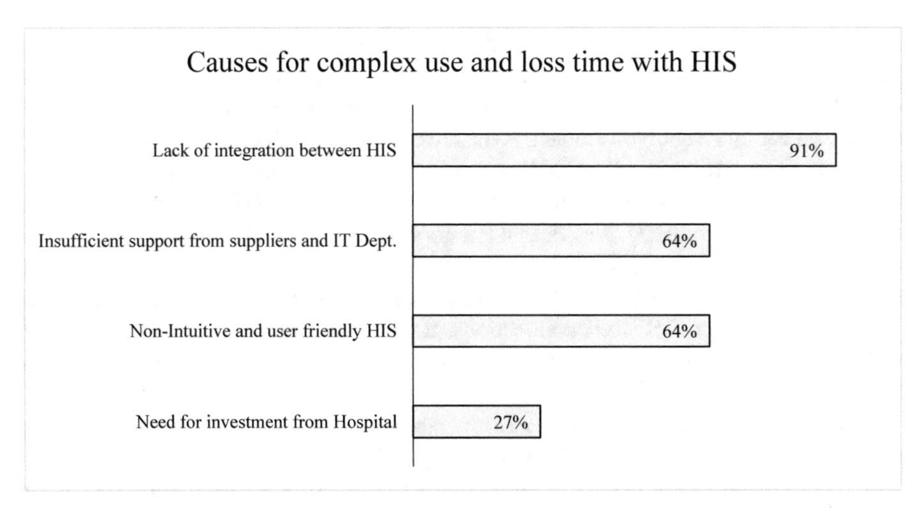

Fig. 4. Causes for complex use and loss time with HIS (Elaborated by the author)

As (Sousa, 2017) pointed there is a complete dependence on the MoH with centralized responsibility for the administrative and healthcare activities processes. The interviewees proved the difficulties existent to establish interconnection between the IS from MoH and the privates EMRs and the lack of an equal patient's identifier (Jardim & Martins, 2016). Public hospitals mentioned in this research have different private IS suppliers which is seen as a difficulty to establish interoperability connection with systems from Company I. According to (de Almeida Simoes et al., 2017) there is a diversity of HIT systems across the Portuguese hospitals and the problems inherent to the healthcare data transmission between the systems and their data storage.

Concerning the healthcare routines that changed with the introduction of IS, all the persons interviewed considered that implementation of IS resulted in the dematerialization of processes. Most of the interviewed persons pointed out that as consequence there was an increase of information access and an efficiency analysis from the data storage plus was also mentioned the reduction of paper flow in the healthcare routines. This is in agreement with the authors (Reza et al., 2020) and (Kruse et al., 2018) that mentioned the increase in the quality of data management, reduction in the paper process, and more available information to the public health policies and research.

However, 94% of the sample interviewed consider the use of this technology is complex and a loss of time and 91% consider the lack of interoperability between health information systems as the cause. Also, (Samal et al., 2016) mentioned that the lack of communication between different IS caused a loss of time in the processes of the health providers. Additionally, the members of the hospital and supplier's company pointed out the fact the HIS is not intuitive and user-friendly without being adapted to the healthcare professionals' routines. Hence, a poor user interface design forces the healthcare professional to switch mental models to interact with the EHR which enhances potential errors in data input and comprehension (Sittig et al., 2020).

Taking into consideration the (da Silva, 2017) conclusions that mention SClínico has a HIS that allows the optimization of clinical information, contributes to support the user's experience, and provides more effective and efficient performance. The data collected is not in total agreement with these conclusions since the healthcare professionals and suppliers that worked with this platform pointed out difficulties in the use. The reason for this negative feedback is the lack of interoperability between this central registration system and the other EMRs on the hospitals as mentioned by (Jardim & Martins, 2016) as the major difficulty faced among the IS operating in Portuguese healthcare units. Further, this research shows that interoperability is conditioned by the type of standards defined between the information systems. The hospital that operates based on HL7 standards presents more the potential of IS and their intercommunication than the hospitals that had DBLINK connections. It would be important to make a detailed analysis of the impact on the efficiency of healthcare routines based on interoperability standards.

Relating the elements to achieve interoperability, healthcare professionals are a key element to have a successful implementation of HIS and the interconnection between them. They are the main users of the EMRs (Furukawa & Pollack, 2020), and a successful HIS implementation will result in healthcare professionals being more satisfied (Davis, 2019). The interviews converge for this conclusion since 91% of the research sample considers the resistance of healthcare providers as a cause for the difficulty faced by the hospitals to achieve the digitalization of processes. In detail, healthcare providers create some resistance and do not see value in working with HIS since they have an overload of tasks to perform in these systems and face difficulties when using them. Taking into consideration the analysis of (Aceto et al., 2018) and this data, it is possible to identify the reduced technological skills of healthcare providers aligned with the lack of interoperability standards between the HIS contribute to the complex use of the healthcare platforms.

Beyond that was also pointed the changing process has a decisive moment to the success of a new system and therefore necessary to evolve the healthcare providers on this process. Hence, the hospital board must have the capability of creating motivation and involving employees during the change process. (Marto, 2017). The data collected shows that 64% of interviewed persons mentioned that is needed an increase in the presence of the IT department and HIS suppliers in the hospital activities since their current support is insufficient. To conclude the key element can be summarized as the promotion of an organizational culture focused on the technological development of health services.

According to the data collected and (Samal, 2016), the use of HIS is complex and translates into a loss of time due to the lack of intercommunication between the systems. It was recognized during the interviews that users complain about the high number of clicks to execute their digital tasks and mentioned that HIS is not intuitive and user-friendly. Due to the poor interface of HIS, it hinders the healthcare professionals' daily routine and enhances potential errors (Sittig, 2020). These errors can be medication and diagnosis lapses (Poly et al., 2018) which can potentially harm or even lead to the death of the patient (Zhou et al., 2018) and be emotionally devastating to the healthcare professional (Mota, D; Moreira de Sousa, A; Ribeiro, 2020). The interview statements are also in concordance with (Vieira, 2018) since the healthcare providers especially nurses that operate with SClínico have the perception of slow navigation and unfriendly structure with a high number of steps. The fact nurses are the healthcare providers that have the worst feedback over the HIS use is due to the time spent executing their daily tasks in SClínico and other EHRs (Bailas, 2016).

5 Conclusion

The study verifies that the lack of interoperability between HIS results in slow and complex use of these systems. Therefore, the healthcare professionals' routine is affected due to time-loss executing their daily tasks. According to (Sittig et al., 2020), this issue can also result in mistakes such as incorrect data input and comprehension, or even duplicated blood tests, wrong medication prescription, or delayed treatments (Coiera et al., 2016) (Poly et al., 2018). These problems demonstrate an absence of a digital health mentality and management focused on the sustainability of systems and software.

The analysis of the Portuguese hospitals exposed the difference between the public and private hospitals' strategies in investment and managing the technological evolution. The public hospital structures are unadapted to the HIS due to the low budgets to invest in technology and the high number of suppliers. On the other side, private hospitals have a higher level of integration between the systems and invest more in IT structures and equipment to guarantee better systems performance.

Healthcare professionals are the main users of the HIS and play a key role in the implementation and management of this technology. This study shows that resistance of healthcare providers is one of the causes to hinder the systems' implementation, especially among the physicians which can be explained by the lack of training and inexperience compared with other healthcare professionals. As a solution to this issue, healthcare institutions should stimulate an organizational culture focused on technology and train physicians and other professionals to support digital change.

The starting point is to understand that interoperability does not depend exclusively on the standards used to interconnect the information systems. Hospitals' structures should prepare for the changes in healthcare routine, including the organizational culture, trained professionals to work with these systems, and a suitable technological structure. In addition, suppliers should cooperate closely with the hospitals to guarantee more functional systems and reduce the loss of time executing tasks.

This research was limited to the COVID-19 pandemic that did not allow physical observation in hospitals and limited the range of persons to participate in this research. Even though the sample allowed to create the previous conclusions and exposed new issues to investigate, it is not legitimate to establish general remarks about the Portuguese national health system. Furthermore, the study's remarks show the need for further investigation about this theme and should be carefully analysed due to the investigation limitations.

References

Aceto, G., Persico, V., Pescapé, A.: The role of information and communication technologies in healthcare: taxonomies, perspectives, and challenges. J. Netw. Comput. Appl. **107**, 125–154 (2018). https://doi.org/10.1016/j.jnca.2018.02.008

Adler-Milstein, J., Holmgren, A.J., Kralovec, P., Worzala, C., Searcy, T., Patel, V.: Electronic health record adoption in US hospitals: the emergence of a digital "advanced use" divide. J. Am. Med. Inform. Assoc. **24**(6), 1142–1148 (2017). https://doi.org/10.1093/jamia/ocx080
Bailas, C.M.: Impacto do uso de sistemas de informação informatizados na carga global de trabalho dos enfermeiros (2016)
Coiera, E., Ash, J., Berg, M.: The unintended consequences of health information technology revisited. Yearb. Med. Informatics **25**(1), 163–169 (2016). https://doi.org/10.15265/iy-2016-014
Venters, C.C., et al.: Software sustainability: research and practice from a software architecture viewpoint. J. Syst. Softw. **138**, 174–188 (2018). https://doi.org/10.1016/j.jss.2017.12.026
da Silva, G.C.S.F: Ferramenta de registo de doenças respiratórias em cuidados de saúde primários. Universidade do Porto (2017)
Davis, T.: Are EMRAM Stage 7 Physicians More Successful? (2019). https://klasresearch.com/resources/blogs/2019/03/29/are-emram-stage-7-physicians-more-successful
de Almeida Simoes, J., Augusto, G.F., Fronteira, I., Hernandez-Quevedo, C.: . Health systems in transition **19**(2) (2017)
Dobrzykowski, D.D., Tarafdar, M.: Understanding information exchange in healthcare operations: evidence from hospitals and patients. J. Oper. Manag. **36**, 201–214 (2015). https://doi.org/10.1016/j.jom.2014.12.003
Dobrzykowski, D.D., Tarafdar, M.: Linking electronic medical records use to physicians' performance: a contextual analysis. Decis. Sci. **48**, 7–38 (2017). https://doi.org/10.1111/deci.12219
Dodd, J.C.: Steps needed to plan and design a patient–consumer-driven architecture. In: Healthcare IT Transformation, pp. 34–53. Taylor & Francis Group (2017)
Jamoom, E.W., Heisey-Grove, D., Yang, N., Scanlon, P.: Physician opinions about EHR use by EHR experience and by whether the practice had optimized its EHR use. J. Health Med. Inform. **7**, 1000240 (2016). https://doi.org/10.4172/2157-7420.1000240
Fan, K., Wang, S., Ren, Y., Li, H., Yang, Y.: MedBlock: efficient and secure medical data sharing via blockchain. J. Med. Syst. **42**(8), 1–11 (2018). https://doi.org/10.1007/s10916-018-0993-7
Frisse, M.E.: Chapter 5 - Interoperability. Key Advances in Clinical Informatics, 69–77 (2017). https://doi.org/10.1016/B978-0-12-809523-2.00005-4
Furukawa, M., Pollack, E.: Achieving HIMSS Stage 7 designation for EMR adoption. Nurs. Manage. **51**, 10–12 (2020). https://doi.org/10.1097/01.NUMA.0000617044.57943.e1
Jardim, S.V.B., Martins, A.C.: An overview and a future perspective in health information systems in Portugal. Encyclopedia of E-Health and Telemedicine, 987–997 (2016). https://doi.org/10.4018/978-1-4666-9978-6.ch077
Kruse, C.S., Stein, A., Thomas, H., Kaur, H.: The use of electronic health records to support population health: a systematic review of the literature. J. Med. Syst. **42**(11), 1–16 (2018). https://doi.org/10.1007/s10916-018-1075-6
Kuo, J.W.Y., Kuo, A.M.H.: Integration of health information systems using HL7: a case study. Stud. Health Technol. Inform. **234**, 188–194 (2017). https://doi.org/10.3233/978-1-61499-742-9-188
LeSueur, D.: reasons healthcare data is unique and difficult to measure. Health Catalyst, 1–5 (2017)
Longo, D.L., Drazen, J.M.: Data sharing. N. Engl. J. Med. **374**, 276–277 (2016)
Martins, C., Duarte, J., Portela, C., Santos, M.: Improving the use of the electronic health record using an online documentation manual and its acceptance through technology acceptance model. In: ICT4AWE 2019 - Proceedings of the 5th International Conference on Information and Communication Technologies for Ageing Well and e-Health, pp. 346–351 (2019). https://doi.org/10.5220/0007878103460351

Marto, V.: A Gestão da Mudança em Sistemas de Informação: a migração do sistema de gestão de doentes para a aplicação SONHO V2 no Centro Hospitalar de Leiria, EPE. Instituto Politécnico de Leiria, Universidade do Porto (2017)

Mota, D., Moreira de Sousa, A., Ribeiro, L.: The emotional impact of medical error on Portuguese medical residents - an exploratory study. European Journal of Public Health **30** (2020). https://doi.org/10.1093/eurpub/ckaa166.1065

Oemig, F., Snelick, R.: Healthcare Interoperability Standards Compliance Handbook. In: Healthcare Interoperability Standards Compliance Handbook (2016). https://doi.org/10.1007/978-3-319-44839-8

Mohaimenul Islam, Md., Poly, T.N., Li, Y.-C.: Recent advancement of clinical information systems: opportunities and challenges. Yearb. Med. Informatics **27**(01), 083–090 (2018). https://doi.org/10.1055/s-0038-1667075

Reza, F., Prieto, J.T., Julien, S.P.: Electronic health records: origination, adoption, and progression. In: Magnuson, J.A., Dixon, B.E. (eds.) Public Health Informatics and Information Systems. HI, pp. 183–201. Springer, Cham (2020). https://doi.org/10.1007/978-3-030-41215-9_11

Salomi, M.J.A., Maciel, R.F.: Document management and process automation in a paperless healthcare institution. Technol. Invest. **08**(03), 167–178 (2017). https://doi.org/10.4236/ti.2017.83015

Samal, L., et al.: Care coordination gaps due to lack of interoperability in the United States: a qualitative study and literature review. BMC Health Serv. Res. **16**(1), 1–9 (2016). https://doi.org/10.1186/s12913-016-1373-y

Sittig, D.F., et al.: Current challenges in health information technology–related patient safety. Health Inform. J. **26**, 181–189 (2020). https://doi.org/10.1177/1460458218814893

Sousa, C.: Excelência da prática clínica evidenciada pelo processo clínico eletrónico, mito ou realidade? Practice of clinical excellence evidenced by EMR, myth or reality? Revista Clínica Do Hospital Prof Doutor Fernando Fonseca **4**, 20 (2017)

Stavert-Dobson, A., Risk, M.C.: HIT impact on patient safety. Health Information Systems, Managing Clinical Risk (2018). https://doi.org/10.1007/978-3-319-26612-1

van Poelgeest, R., et al.: Level of digitization in Dutch hospitals and the lengths of stay of patients with colorectal cancer. J. Med. Syst. **41**(5), 1–7 (2017). https://doi.org/10.1007/s10916-017-0734-3

Vieira, S.M.C.: Use and Evolution of Nursing Information Systems: Influence on Decision Making and Quality of Nursing Care. Universidade do Minho (2018)

World Health Organization: From Innovation to Implementation eHealth in the WHO European Region (2016)

Asi, Y.M., Williams, C.: The role of digital health in making progress toward sustainable development goal (SDG) 3 in conflict-affected populations. Int. J. Med. Informatics **114**, 114–120, ISSN 1386-5056 (2018). https://doi.org/10.1016/j.ijmedinf.2017.11.003

Zhou, S., Kang, H., Yao, B., Gong, Y.: Analyzing medication error reports in clinical settings: an automated Pipeline approach. In: AMIA Annual Symposium Proceedings, AMIA Symposium, vol. 2018, pp. 1611–1620 (2018)

Self-employment as a Response to the Great Resignation

Gustavo Morales-Alonso(✉)

Department of Industrial Engineering, Business Administration and Statistics, Universidad Politécnica de Madrid, Madrid, Spain
gustavo.morales@upm.es

Abstract. A Great Resignation is going on. Workers are leaving their jobs at rates never seen before. They do it under economic circumstances never experienced by a whole generation of workers: the highest inflation rates in more than 40 years, in presence of the longest period of low interest rates and with apparently overvalued stocks and real estate markets. Not surprisingly, in such scenario political discontent arises, leading to an unprecedented success of populists' proposals at political elections. To this date, mainstream theories at economic, business management and political levels appear unable to fix the multiple existing economic and social problems. In this paper, it is argued that the use of information and communication technologies provides with new perspectives for business management, in terms of relocation of workers at different geographic settings, but also establishing different contractual situations with their (former) employers and performing different functional-related tasks at the value chain level. That is, uncertain times bring new opportunities for younger workers, who may opt to live as digital nomads, locating their residence where less taxes need to be paid. Renouncing to work for a company, they can establish their selves as freelancers, self-employed or entrepreneurs, with the aim of offering their services to companies or individuals located anywhere, while developing a more nuanced knowledge of companies and industries. Developing these new business management perspectives requires, however, a deeper understanding of how entrepreneurial intentions relate to other factors, such as political positioning, human values or religious feelings.

Keywords: Entrepreneurship · Inflation · Free-lancing

1 Introduction

A generation of young workers is now experiencing the curse "may you live in interesting times". The appearance in 2021 of the Great Resignation, the label given to the voluntary resignation of millions of workers in all parts of the world, made it clearer than ever that we live in interesting times. Not in vain, the rise of populist regimes can drive the world to a dead end, in which even democracy can be at risk. If we are to defend the way we live, we need more technology, more social collaboration, more trade, and more free markets, and not less. The dynamics that have led to this situation, and the policies to solve it, are a call to arms to the academic community.

V. Gupta et al. (Eds.): SSEBIM 2022, LNISO 62, pp. 122–130, 2023.
https://doi.org/10.1007/978-3-031-32436-9_10

The abovementioned interesting times are new for a generation of workers, but are not entirely new for the world. In fact, the decade of 1970 was also very interesting, for the bad reasons. Nevertheless, in the mid-1980s, the so-called Great Moderation was established. It consisted of a series of imposition of economic measures, altogether with business administration and public policies, in a favorable environment for it. It lasted until the financial crisis of 2008 and brought unprecedented prosperity.

Bernanke (2004) attributes the success of the Great Moderation to three factors: (1) structural change, (2) improved macroeconomic policies and lastly, (3) luck, since during those years there were no major shocks from the environment. The decades of the Great Moderation have been of a great economic and social prosperities, attributed to the free markets, with minimum public intervention in those sectors that are considered sensitive, and a balanced collection of taxes as a way to finance the welfare state. This has been possible due to the existence of convenient demography in developed countries, with an adequate population pyramid.

Last but not least, the environment helped. First, with the fall of the Berlin wall, the enemies of trade were definitively eliminated, with the consequent increase in the size of global markets, which led to the beginning of Globalization. Second, with the prolonged period of peace that followed the Cold War. Third, with the breakthrough efforts performed by the countries of the Eurozone to promote trade, such as the single European market (1993), the Schengen agreement (1995), the creation of the European Central Bank (1998) or the implementation of the Euro (1999). Fourth, the charismatic leadership exercised by leaders such as Margaret Thatcher (1979–1990), Ronald Reagan (1981–1989), Helmut Kohl (1982–1998) or Pope John Paul II (1978–2005), to name a few.

2 A Problem of Political Economy

The Great Moderation is anchored on the monetarist theories of Milton Friedman, toned down with actions of a fiscal nature proposed by neo-Keynesians such as Joseph Stiglitz or Paul Krugman, while companies put into practice the business management theories developed by Michael Porter. This, on the one hand, has made it possible to establish a series of balances between monetary policy and fiscal policy, preventing excesses from being committed either from the fiscal point of view or from the monetary point of view, giving rise to sustained growth over time. On the other hand, the great innovations that have occurred in the field of business management have allowed increases in productivity to be the foundations of healthy economic growth, which has reduced poverty in the world by billions of people.

But there is no economic growth that can last a lifetime. In fact, the growth led by the Great Moderation found its limit in the financial crisis of 2008, in which deregulation and the irrational exuberance of the markets ended up creating financial bubbles, especially in the real estate markets of different countries. The success of the macroeconomic policies of the Great Moderation gave credence to the new macroeconomic proposals that sprang from 2008, which have been eminently expansionary in both fiscal and monetary policies.

From the fiscal point of view, for example, although the European Monetary Union promoted that the member countries should not exceed 3% of the annual deficit and that the total accumulated debt should not exceed 60% of the GDP of each country, the reality is that this has been systematically breached to the point where Germany, Spain and Italy presented structural budget balances of −3.1%, −5.3% and −5.9% of GDP for the year 2020, respectively. This causes these three countries to present a global debt over their GDP for the year 2021 of 69%, 119% and 156%, respectively. The United States presented a deficit of −10.7% in that year, and an accumulated debt of 137% of its GDP (IMF, 2021).

From the monetary point of view, unusually low interest rates have been allowed by both the US Federal Reserve and the European Central Bank and the Bank of England. These very low interest rates have flooded the market with liquidity, allowing very easy access to credit, and have sent the wrong signals to economic agents. These wrong signals refer to the fact of undertaking investments that were only good due to their very low cost of financing, but that would never have been undertaken if there had been a higher cost of financing.

It should be noted here that the main evaluation metric of fiscal and monetary policies is precisely the level of inflation detected in the economy. And given that global inflation rates did not exceed 2% per year, the coordination of expansionary fiscal and monetary measures seemed to be safe.

It is in this environment that a shock event occurs, attributable to the third group of those mentioned by Bernanke (2004): bad luck. Also identifiable as an unpredictable event, a black swan as defined by Taleb (2007), the COVID-19 pandemic. This pandemic, and especially the confinements to which it gave rise, motivated the countries of many governments, as well as the monetary authorities, to redouble their efforts in the same direction in which they had been making them: more expansionary fiscal policy and more expansionary monetary policy.

The monumental public debts contracted by various countries have already been discussed on this paper. But to understand the level of monetary excesses that have occurred, it should be highlighted that, of the total of 6,000 billion dollars that had been printed throughout history (until December 2021): 1,000 billion were printed in the 48 years from 1960 to 2008; 3,000 billion were printed in the 6 years from 2008 to 2014; and after a period of calm, another 2,000 billion dollars were printed in less than 2 years, between 2020 and 2021. In short, since 2008, 5,000 billion dollars have been printed, 80% of the total amount that has existed since the first dollar was printed in 1792 (FED, 2022).

In the year 2021, the policies that seemed to have been working in the previous fifteen years began to show signs of exhaustion, evidenced by the rise in inflation rates. This unusual rise in inflation may be due to (1) expansionary fiscal policies that have put more money than normal into the hands of economic agents, fueling their demand for goods and services, which has not been met by supply; (2) a paradigm shift in energy markets, in which discourses in favor of renewable energy have reduced investment in more traditional energy sources (gas, hydrocarbons, nuclear power), at a time when they are indispensable to maintain the economic structure of developed countries; (3) the expansive monetary policies that have been carried out, which have given rise to "too much money in search of too few goods", that is, the money available in the economy has grown more than the goods available to be bought with that money, increasing the value of the goods; and (4) the existence of bottlenecks in the supply chains, which have been interrupted by the mandatory lockdowns. The savings of families during the confinements would have fueled the demand for goods and services as these confinements ended, finding supply chains stressed and without the possibility of satisfying that demand for goods and services.

These four events are not mutually exclusive, they can occur simultaneously. Moreover, in addition to occurring intermingled, they may have different levels of importance in different parts of the world, which would mean that the measures that can solve inflation may not be the same in all countries.

3 The Great Resignation

As if that were not enough, in this environment of high levels of inflation a new factor appears that comes to introduce even more entropy into the system: the Great Resignation, a process by which the turnover of workers in certain industries begins to increase in a very remarkable way. More than 24 million US employees left their jobs in 2021 (Sull et al., 2022). Coined by Anthony Klotz of Texas A&M University in 2021, this term denotes a process that has been detected first in the United States, where more than 40% of all employees were thinking of leaving their jobs at the beginning of 2021, triggering a mass resignation process as the year progressed.

There are many reasons for the Great Resignation. In fact, as many as people who have resigned. Thus, if all happy families are alike, but the unhappy ones are each in their own way, it can be said that the satisfied workers are all for similar reasons, but the dissatisfied ones each have their own particular reason. The reasons for the lack of job satisfaction stem from both economic and non-economic causes. In turn, within the economic causes, these may have different origins, evidencing the complexity of the process.

Among the plausible causes that have been pointed out, it is worth highlighting the change in professional values due to the psychological shock of the lockdowns and the pandemic, the desire not to return to the office for health reasons or to avoid long hours of commuting, the existence of toxic work cultures incapable of promoting diversity, equality and inclusion, the feeling of job insecurity, the high levels of innovation of certain companies, which generate very demanding work environments, or the inability of companies to recognize the merits of employees (Serenko, 2022; Sull et al., 2022).

The Great Resignation is important to the extent that millions of people make the same decision in a relatively short space of time. But, in the author's opinion, Google's historical difficulty in retaining its workers shows that this is not an entirely new phenomenon, but that it has increased in scale due to shocks from the environment. Indeed, Google is known for keeping a turnover of workers at a very low level, 1.1 years, failing even in retaining some of its workers who have contributed to developments such as Google Maps and Google Wave (Lars and Jens Rasmussen) or Google Meet (William Wen). In essence, the rise of information and communication technologies make possible the appearance of new forms of work, among which remote work stands out, allowing the existence of digital nomads. The existence of shocks in the environment mobilize psychological traits that lead a large number of workers to consider the development of portfolio careers, in which the worker's tasks are not fixed, but flow, allowing him to go through various positions within a company, various companies, or various contractual situations. Furthermore, and not least, the free movement of workers within certain economic zones allows the relocation of workers, who flee from abusive and confiscatory tax structures, letting them at the same time to obtain non-monetary but experiential enrichment through their changes of residence.

It is in this context, in which has to be considered to what extent self-employment can be an escape route for workers who are within the Great Resignation trend. This is because working as a self-employed, entrepreneur or freelancer can answer many of the concerns that are behind the set of massive job resignations that are being experienced. But it certainly generates other kinds of concerns to which an employee is rarely exposed. In this respect, the business management literature is used to point out a series of underlying factors that suggest greater success at the individual level in pursuing a professional career as a self-employed person.

4 Cognitive Traits in the Preference for Self-employment

4.1 The Entrepreneurial Intention

The last few decades have been rich in scientific literature regarding the drivers that lead some people to start businesses when most do not. A large number of factors have been proposed as drivers of business action, such as those presented in the conceptual framework of the Global Entrepreneurship Monitor (GEM) (Bosma et al., 2020). In it, contextual factors, social values about entrepreneurship and individual attributes are coordinated with each other, acting as triggers for entrepreneurial activity. Different researchers have presented results regarding the prevalence between these factors, such as Pinillos and Reyes (2011) or Morales-Alonso et al., (2020, 2022), evidencing the complexity of the relationships between them.

Entrepreneurial behavior is volitional and driven by cognitive mechanisms (Morales-Alonso, et al., 2015), and therefore the triggering factor of entrepreneurial actions is the development of entrepreneurial intentions (Ajzen, 1991). This relativizes the importance of contextual factors, social values and individual attributes (Krueger et al., 2000). Intention can be defined as the trigger for actions and behaviors. For this reason, the decision to found a company is defined as an intentional process in which the individual will make a reasoned decision, with the support of their immediate environment and with the conviction that they have the necessary knowledge to achieve success, based on what he/she considers to be an interesting market niche. Hence, the entrepreneurial spirit is considered a planned behavior, and can be predicted through intention models (Ajzen, 1991; Krueger et al., 2000).

4.2 Human Values and Personality

Nineteenth-century German sociology, of which Max Weber is its greatest exponent, argues that our relationship with other people is what defines our behavior, since humanity is a social species. The exchanges between people favor the creation of a human value or culture, that is, a set of beliefs, norms, habits, values, assumptions, symbols and rituals that have a strong influence on individual attitudes, intentions and behaviors (Hofstede, 1980; Schwartz & Bilsky, 1987).

Within these human values, individualism or autonomy is found, which refers to the tendency to guide values and actions towards independence and the achievement of personal or group interests (Hofstede, 1980; Schwartz, 1999). The human value that opposes this is collectivism or conservatism, which is characterized as the inclination to act cooperatively in the interest of the group, taking into account the expectations of the immediate group (Hofstede, 1980; Schwartz, 1999).

Entrepreneurial action is strongly related to individualistic values (Morales-Alonso et al., 2020; Pinillos & Reyes, 2011), since the autonomy represented by these people makes them more likely to get involved in less predictable situations than collectivist individuals tend to do perceive as extremely risky.

A second human value closely associated with business action is mastery-harmony as in Schwartz (1999), which corresponds to masculinity-femininity in terms of Hofstede (1980). According to this value, masculine individuals are characterized as being competitive, ambitious, or achievement seekers as opposed to feminine individuals who are more concerned with their interpersonal relationships and a general feeling of harmony or need to fit in with the world.

Finally, a third value frequently related to entrepreneurial behavior is the aversion to uncertainty, which comes from the aversion to risk present in a society. That is, it measures the extent to which members of a society tolerate ambiguity and are confident in making decisions in an uncertain environment.

4.3 Libertarianism vs. Estatism

After the fall of the Berlin wall, and with it, the regimes that represented greater intervention in the economy, it became clear that economic development is closely linked to economic freedom. The Spanish thinker Antonio Escohotado argued that "generically, capitalism needs democracy more than democracy needs capitalism". Thus, if we define democracy as the political system that defends the sovereignty of the people and the right of the people to elect and control their rulers, we will understand that innovation and progress go hand in hand with inclusive political systems that limit the power of elites. Extractive, as argued by Acemoglu and Robinson (2012).

Boudreaux et al. (2019) state that those people who are prone to entrepreneurial venture and who reside in environments that enjoy higher levels of economic freedom, have a greater probability of creating a company. In this way, economic freedom is linked to business action and precedes economic growth, as shown by McMullen et al., (2008).

Conversely, the perception of economic-administrative barriers that hinder the creation of new companies, such as financial and credit difficulties, administrative obstacles and lack of institutional support, play a significant negative role in entrepreneurial behavior (Lüthje & Franke, 2003).

In short, the existence of contextual or institutional frameworks, as well as the perception that individuals have of them, makes certain contexts more conducive to entrepreneurship than others. There will be less propensity for private investment and risk taking in environments where the risks of confiscation are higher. Among the existing risks, the lack of legal equality and adequate regulation, unstable monetary conditions, instability in institutions and criminality, among others, can be highlighted (Espinosa et al., 2021).

According to these arguments, economic development is only possible if property rights are guaranteed and if these rights are violated, offenders are sanctioned and prosecuted (Espinosa et al., 2022; Wang et al., 2021).

4.4 The Religious Beliefs

Religious sentiment also dictates behavioral norms that affect both believers and non-believers through the values that religion arouses (Adamczyk & Palmer, 2008). In many disciplines, the influence of religious feelings has been underestimated, despite the fact that more than 80% of the world's population professes some religion, divided mainly into the four most common, namely, Christianity (32%), Islam (23%), Hinduism (16%) and Buddhism (7%) (Pew Research Forum, 2010).

Religion is defined by Iannaccone (1998) as a shared set of beliefs, activities and institutions based on faith in supernatural forces, which is strikingly similar to Hofstede's (1980) definition of national culture. Parboteeah et al., (2015) maintain that religiosity goes "above and beyond" culture because religion indicates what is acceptable and what is not acceptable in human behavior.

Regarding entrepreneurship, we can consider that religion acts as a complement to the definition of culture or human values, as suggested by Farmaki et al., (2020). Not in vain, different researchers highlight the importance of religious sentiments in relation to self-employment, for example, in Audretsch et al., (2013), Parboteeah et al., (2015), Smith and McMullen (2021) or Weber (1930).

Weber (1930) was the strong advocate of Protestantism values fostering entrepreneurial attitudes, capitalism, and development. More recently, Nunziata and Rocco (2016) have argued that Protestantism encourages entrepreneurship among religious minorities, although no evidence of this was found when comparing religious majorities.

5 Discussion, Conclusion and Expectations for the Future

The only alternative to trade is violence. This inspirational quote of Spanish thinker Antonio Escohotado highlists the risks of establishing political regimes in which the social collaboration based on free trade is limited.

The main discussion today at economic, social and political level is not between left- and right-winded policies, but rather between supporters of the free markets and supporters of state intervention. The thesis of this paper is that existing welfare states in developed countries are not sustainable with the current age distribution of tax payers, which leads to permanent budgetary deficits that build up monstrous debts at country level. Monetary institutions have kept low interest rates for decades, with the aim of allowing these debts to be paid. But when the high inflation rates have become the norm, monetary policies to reduce supply of money have begun, making it impossible for the highly indebted countries to cope with their debt.

In this scenario, populist discourses become very attractive for voters (tax payers). Blaming the rich, the markets, or the financial system is an easy way to obtain political power. But it will probably not suffice to retain it on a democratic system. The only way in which populism can hold political authority in a timely manner is moving towards anti-democratic structures. Germany in 1930 and Venezuela in 2000 provide good examples of this.

The Great Resignation speaks, at a business management level, of employees (tax payers) discontent with the current state of affairs. This weariness is rooted on the feeling of that policies that aimed to improve economic and social perspective do not succeed. The idea of a social contract existing between the individual and the society where he/she lives starts to be questioned, which can eventually lead to inter-generational conflicts in developed countries.

There is no room for despair. The future is not to be waited for, it is to be built up. The process of social cooperation is the way in which the human action, from a Misean and Kirznerian perspectives, takes place in search of a more satisfactory situation for the individual. As a byproduct of this, free markets produce economic wellbeing, the cornerstone of social content.

In this regard, the academic community has a responsibility. More research is needed for a deeper understanding of how the search for economic freedom among the generation of younger workers is intertwined with their human values, religious feelings and political inclinations. Knowing these is indispensable for the advent of economic measures, business management techniques and public policies that can succeed in the future.

References

Acemoglu, D., Robinson, J.A.: Why Nations Fail: The Origins of Power, Prosperity and Poverty, 1st edn. Crown, New York (2012)

Adamczyk, A., Palmer, I.: Religion and initiation into marijuana use: the deterring role of religious friends. J. Drug Issues **38**(3), 717–741 (2008). https://doi.org/10.1177/002204260803800304

Ajzen, I.: The theory of planned behavior. Organ. Behav. Hum. Decis. Process. **50**, 179–211 (1991). https://doi.org/10.1016/0749-5978(91)90020-T

Audretsch, D.B., Bönte, W., Tamvada, J.P.: Religion, social class, and entrepreneurial choice. J. Bus. Ventur. **28**(6), 774–789 (2013). https://doi.org/10.1016/j.jbusvent.2013.06.002

Bernanke, B.: The Great Moderation. Washington, DC (2004)

Bosma, N., Hill, S., Ionescu-Somers, A., Kelley, D., Levie, J., Tarnawa, A.: Global Entrepreneurship Monitor 2019/2020 Global Report, GEM Publications (2020). https://www.gemconsortium.org/report/gem-2019-2020-global-report, last accessed 2022/09/27

Boudreaux, C.J., Nikolaev, B.N., Klein, P.: Socio-cognitive traits and entrepreneurship: the moderating role of economic institutions. J. Bus. Ventur. **34**(1), 178–196 (2019)

Espinosa, V.I., Alonso Neira, M.A., de Soto, J.H.: Principles of sustainable economic growth and development: a call to action in a post-covid-19 world. Sustainability (Switzerland) **13**(23), 1–14 (2021). https://doi.org/10.3390/su132313126

Espinosa, V.I., Wang, W.H., de Soto, J.H.: Principles of nudging and boosting: steering or empowering decision-making for behavioral development economics. Sustainability (Switzerland) **14**(4), 1–18 (2022). https://doi.org/10.3390/su14042145

Farmaki, A., Altinay, L., Christou, P., Kenebayeva, A.: Religion and entrepreneurship in hospitality and tourism. Int. J. Contemp. Hosp. Manag. **32**(1), 148–172 (2020). https://doi.org/10.1108/IJCHM-02-2019-0185

Federal Reserve (FED): Monetary Base, Total (2022). https://fred.stlouisfed.org/series/BOGMBASE, last accessed 2022/09/27
Hofstede, G.: Culture's Consequences: International Differences and Work Related Values. Sage (1980)
Iannaccone, L.R.: Introduction to the economics of religion. J. Econ. Lit. **36**(3), 1465–1495 (1998)
International Monetary Fund (IMF): World economic outlook - recovering during a pandemic (2021). https://www.imf.org/-/media/Files/Publications/WEO/2021/October/English/text.ashx%0A, last accessed 2022/09/27
Krueger, N.F., Reilly, M.D., Carsrud, A.L.: Competing models of entrepreneurial intentions. J. Bus. Ventur. **15**, 411–432 (2000). https://doi.org/10.1016/S0883-9026(98)00033-0
Lüthje, C., Franke, N.: The "making" of an entrepreneur: testing a model of entrepreneurial intent among engineering students at MIT. R and D Management **33**, 135–147 (2003). https://doi.org/10.1111/1467-9310.00288
McMullen, J.S., Bagby, D.R., Palich, L.E.: Economic freedom and the motivation to engage in entrepreneurial action. Entrep. Theory Pract. **32**(5), 875–895 (2008). https://doi.org/10.1111/j.1540-6520.2008.00260.x
Morales-Alonso, G., Pablo-Lerchundi, I., Núñez-Del-Río, M.C.: Entrepreneurial intention of engineering students and associated influence of contextual factors / Intención emprendedora de los estudiantes de ingeniería e influencia de factores contextuales. Revista de Psicología Social **4748** (2015). https://doi.org/10.1080/02134748.2015.1101314
Morales-Alonso, G., Nuñez Guerrero, Y., Aguilera, J.F., Rodrıguez-Monroy, C.: Entrepreneurial aspirations: economic development, inequalities and cultural values. Eur. J. Innov. Manag. **24**, 553–571 (2020). https://doi.org/10.1108/EJIM-07-2019-0206
Morales-Alonso, G., Blanco-Serrano, J.A., Guerrero, Y.N., Grijalvo, M., Jimenez, F.J.B.: Theory of planned behavior and GEM framework–how can cognitive traits for entrepreneurship be used by incubators and accelerators? Eur. J. Innov. Manag. (ahead-of-print) (2022)
Nunziata, L., Rocco, L.: A tale of minorities: evidence on religious ethics and entrepreneurship. J. Econ. Growth **21**(2), 189–224 (2016). https://doi.org/10.1007/s10887-015-9123-2
Parboteeah, K.P., Walter, S.G., Block, J.H.: When does christian religion matter for entrepreneurial activity? The contingent effect of a country's investments into knowledge. J. Bus. Ethics **130**(2), 447–465 (2015). https://doi.org/10.1007/s10551-014-2239-z
Pew Research Forum.: US Religious Knowledge Survey. Washington, DC (2010)
Pinillos, M.-J., Reyes, L.: Relationship between individualist–collectivist culture and entrepreneurial activity: evidence from global entrepreneurship monitor data. Small Bus. Econ. **37**(1), 23–37 (2011). https://doi.org/10.1007/s11187-009-9230-6
Smith, B.R., McMullen, J.S., Cardon, M.S.: Toward a theological turn in entrepreneurship: how religion could enable transformative research in our field. J. Bus. Ventur. **36**(5), 106139 (2021)
Schwartz, S.H.: A theory of cultural values and some implications for work. Appl. Psychol. **48**(1), 23–47 (1999). https://doi.org/10.1111/j.1464-0597.1999.tb00047.x
Schwartz, S.H., Bilsky, W.: Toward a universal psychological structure of human values. J. Pers. Soc. Psychol. **53**(3), 550–562 (1987). https://doi.org/10.1037/0022-3514.53.3.550
Serenko, A.: The great resignation: the great knowledge exodus or the onset of the great knowledge revolution? J. Knowl. Manag. (ahead-of-print) (2022)
Sull, D., Sull, C., Zweig, B.: Toxic culture is driving the great resignation. MIT Sloan Manag. Rev. **63**(2), 1–9 (2022)
Taleb, N.N.: The Black Swan: The Impact of the Highly Improbable, Vol. 2. Random House (2007)
Wang, W.H., Espinosa, V.I., Peña-Ramos, J.A.: Private property rights, dynamic efficiency and economic development: an Austrian reply to neo-Marxist scholars Nieto and Mateo on cyber-communism and market process. Economies **9**(4), 165 (2021). https://doi.org/10.3390/economies9040165
Weber, M.: The Protestan Ethic and the Spirit of Capitalism. Scribner (1930)

Analysis of Information Technology Uses at Construction Projects

Siham Farrag(✉), Yusra A. L. Rashdi, Mohammed Abushammala, and Shalin Prince

Middle East College, Muscat, Oman
siham@mec.edu.om
http://www.mec.edu.om

Abstract. The use of Information Technology (IT) in construction projects is essential for the development of the company's works. This study aims to evaluate the use of IT in the construction industry in the Sultanate of Oman. It also seeks to identify the programmes primarily used in the construction industry, the factors that affect the use of IT, and the approaches to overcoming the barriers to using this technology in construction projects. Initially, an interview was conducted with professional engineers and contractors in the selected construction industry and Ministry of Housing related to implementing IT. Then, an electronic questionnaire was completed, and the data collected from the questionnaire were analysed with the Statistical Package for Social Sciences (SPSS). In general, the analysis showed that all interviewees had a clear understanding of the importance and scope of IT in the construction industry; however, there were still several barriers to its implementation. The results showed that the most common obstacle that reduces the use of IT in the construction industry in the Sultanate of Oman is the weak capabilities of employees in the use of IT. Also, we found that the primary strategy that might help reduce the barriers to using IT in construction projects is to intensify courses in the use of IT. Finally, the results of our study might help the decision-makers in ministries and construction projects in the Sultanate of Oman towards improving the construction sector.

Keywords: Construction Management · Construction Industry · Oman

1 Introduction

The construction field is characterised by a large amount of data, complicated business and contractual relationships, many participants, and the uniqueness of projects [1]. Large construction projects store great information before, after, and during construction on the site. These large projects need Information Management Systems (IMS), data entry, processing, and publishing [2]. Based on the established uniqueness and specifics, it can be stated that effective management is key to the success of not only projects but also construction companies.

Construction management efficiency has become critical not only to fulfill common goals such as project completion within budgeted cost, schedule, and quality, but also to emphasize environmental sustainability for better living and the future generation.

V. Gupta et al. (Eds.): SSEBIM 2022, LNISO 62, pp. 131–141, 2023.
https://doi.org/10.1007/978-3-031-32436-9_11

The country and local governments have released their most recent development plan standards for sustainable development, which entails ensuring that all development is economically, socially, and environmentally sustainable before prior consent is granted [3].

The use of IT is endless, allowing for customization to the demands of the sector for future development dealing with work complexity and sustainable progress [3].

Over the past twenty years, IT has witnessed tremendous innovations, for example, there was an increase in broadband internet subscriptions and mobile subscriptions [3]. Moreover, in the construction industry today, there is a growing trend for IT to be used from conception to completion of a project process [4]; with the development of computers, computer software (e.g., Computer-Aided Design (CAD) and Building Information Modelling (BIM), internet, mobile phones and so on. IT can significantly assist in project planning, organizing, operation and control. The primary use of IT in the construction industry, in general, is office application, computer-aided design (CAD), tools software and communication networks. Implementing IT in office work mainly involves the automation of routine tasks, including the exchange of building documents in digital form [5]. Different types of software applications can perform various tasks:

Cost evaluation software can evaluate construction cost according to the given bill of quantities, Quota management software can manage quota data, Quantity calculation software can calculate amounts according to CAD drawings, and Steel quantity calculation software can calculate steel quantities.

Internet-based communication is the area that is growing the most swiftly. Most construction firms are connected to external networks for information such as material price and cost index [6]. The rapid growth in information management has dramatically impacted business systems. The expansion and integration of IT has been a catalyst among the competing organisations at the global level. Producing professionals in the construction industry will depend on the opportunities arising from the explosion of IT and the reliance on new technologies [7]. Construction projects cannot be improved and developed without the transfer and implementation of computing and other advanced technologies [4].

The construction process, which was previously viewed as ineffective, has now become a significant goal in research projects and programmes in countries worldwide. An example of a programme is the Swedish Research and Development Programme in construction and building. The aim of these activities was the construction products and technology that assisted in developing the construction process. Craig and Sommerville 2006 [2] provide an electronic IMS system that works to save and manage files, which reduces the responsibility of individuals or the institution in preserving and storing files and information. In addition, the IMS also can deal with performance management and reporting systems that not only assist in project management but also in managing the organisation of the whole project. Stephen and David (2006) [8] proposed a framework called PMS-GIS (Progress Monitoring System with Geographical Information Systems) to address development progress regarding a Critical Path Method CPM plan as well as a graphical portrayal of the development synchronised with the plan for getting work done. In PMS-GIS, the structural plan is executed utilising a PC-supported drafting (CAD) program (AutoCAD), the procedure for getting the job done is created using and undertaking the board programming (P3), the plan and timetable data (counting per cent complete data) are connected to a GIS bundle (ArcViewGIS). For each update, the framework delivers a CPM-produced bar outline close by a 3D delivery of the task set apart for progress [8].

The use of information management in the design process makes it possible to evaluate the design and assess the impact of design decisions on project activities [9]. Several studies indicate the benefits of using IT, these include saving time, reducing administrative costs long-term, increasing employee productivity, and improving business management processes. Onyegiri and Nwachukwu (2011) [4] studied the importance, requirements, and obstacles to effectively using ICT in the construction industry. The authors conducted an in-depth analysis of relevant literature. The research mentioned that the significant benefit of using ICT in the construction sector was that employees do not necessarily need to be present at the same venue; computers and the internet allow them to co-ordinate from different locations. Moreover, the requirements mentioned in the study include having basic knowledge of communication technology and establishing a proper code of conduct and regulations by private boards. These factors facilitate the efficient and effective implementation of ICT in the construction sector [4]. Klinc et al. (2016) [10] explored information and communication technology in construction projects. The paper highlighted that a collaborative engineering environment for information sharing is required for successful project delivery. Furthermore, collaborative information sharing has resulted in two diverse communication typologies. Firstly, since information can be accessed anywhere in the world, the communication model has become standardised. Secondly, the users of the information-sharing application could communicate with each other whenever they found it necessary [10].

Up to that end, in current times, the construction industry is becoming highly dependent on the flow of information. Different parties are involved in a single construction project; at any point in time, several professionals will be working in collaboration. To make effective decisions, they must obtain timely and accurate information on their tasks [11]. The rapid development in the Gulf Co-operation Countries highlights the need for countries to understand knowledge management. The government of the Sultanate of Oman, one of the Gulf countries, has launched a development strategy, which is a long-term strategy called Oman Vision 2040, which highlights information and communication technology and the knowledge economy. However, the Omani construction sector, in particular, is falling behind on this vision. There are several constraints and impediments to the effective development of the industry, and there is a large gap between the way the construction sectors operate in advanced economies and those in Oman [12].

However, several studies have been conducted that clarified the use of IT in the construction industry in general, and in the pre-contract stage in particular, and also explained the factors that reduce its use with solutions to these factors in developed countries. Therefore, since it is not possible in developing and underdeveloped countries to implement the solutions that were applied in developed countries, a study needs to be conducted in developing those underdeveloped countries to evaluate the use of IT in the pre-contract stage in construction projects at that context.

Therefore, this study aims to study factors affecting the implementation of IT at the pre-contract stage in construction projects in the sultanate of Oman. This paper focuses on three research objectives. The first research objective is to identify commonly used IT tools in construction companies in Oman. The second objective is to highlight significant barriers hindering IT implementation growth. Finally, the third research objective is to analyse the strategies for mitigating obstacles encountered while attempting to implement IT.

2 Research Methodology

To accomplish the objectives of this study, data was collected through both qualitative and quantitative methods. Initially, in-depth interviews involve intensive individual interviews with a small number of respondents to explore their perspectives on a particular idea, programme or situation [13]. The interview questions were open to allow the respondent free expression with the answer [14]. Furthermore, a web-based survey was used to collect data from our target construction group. Due to the difficulty of obtaining replies from construction company employees in Oman, the questionnaire was e-mailed to 130 contacts in 5 firms and ministry of housing based on the authors' academic and personal contacts. The total number of valid acquired responses was 67.

An interview and questionnaire targeted a construction group of different professions in the Sultanate of Oman. The construction group, included civil engineers, architects and quantity surveyors at the Ministry of Housing and samples of construction companies (5 companies) [15]. The questionnaire contains 25 questions, and it consists of 3 parts that consist of three parts each;

1. Professional details (This section contains questions related to the respondent's profession),
2. Organisational details: (This section contains questions related to the organisation in which the respondent works,
3. Operational details: (This section contains questions about operating software and applications used in construction projects).

Among these questions, there are dichotomous questions such as yes/no, multiple-choice questions, and demographic survey questions [14].

3 Results and Discussion

The interview data was analysed and findings were organised by relevance to the research objectives. Then the data obtained from the survey were analysed using the Statistical Package for Social Sciences (SPSS).

3.1 Interview Analysis

An electronic interview was conducted with five professionals from the construction group of different professions regarding IT in construction, or in the pre-contract stage in particular. Five specific interview questions that include:

- In your opinion, what are the technologies and programs used in your organization which are related to information technology in construction?
- Based on your suggestion, what can be done to boost the level of IT standards in the Oman construction industry?
- Can you talk about the implication of information technology in the Ministry of Housing projects? What exactly is the use?
- What are the factors that prevent using information technology in the pre-contract stage or in construction projects in general? What are the proposed solutions to overcome these factors?
- In what way do you think information and communication technology can aid your professional duties and effect on project sustainability mangement?

The following responses were obtained:

- In general, there are several programs currently used in construction companies and ministries in Oman, such as design programs e.g., AutoCAD, Revit, 3D Max and Photoshop, design appraisal, project management, information storage and retrieval, cost estimation, structural analysis, on-site management, facilities management. Whereas IT is used to store, retrieve and transmit data, the One Drive programme is used to save construction projects for the organisation.
- The respondents reported that using IT helps transfer construction information or bidding processes to other organisations and companies quickly and helps in marketing for construction and keeping projects confidential. It also helps quick communication between engineers and other organisations, gathers real-time data to make quick decisions, and improves employee productivity. In addition, it assists in linking the networks and facilities for large and mega-sized projects.
- According to the responders, implementing web-based project collaboration and developing IT technologies not only meets project budget, schedule, and quality requirements; it also meets functional requirements and client or end-user satisfaction in sustainable development.
- They also listed several factors that may prevent using IT in the pre-contract stage or construction projects in general that included; lack of awareness of the use of IT in the pre-contract stage, a lack of investment and research in the benefits of information and communication technology in construction projects, the weak ability of employees is one of the reasons that hinder the use of IT, lack of sufficient computers in the company, and lack of programmes that are specific to construction. Also, poor training on how to use IT in construction companies.

To that end, the professionals suggest some solutions to overcome these obstacles; such as intensive courses should be conducted in IT to improve the capabilities of employees, and increase the awareness of the importance of using IT in construction companies. Finally, the ministries and the authorities' organisation should establish strict laws on using IT in construction projects.

3.2 Survey Analysis

The analysis of this survey is based on 67 valid responses from professionals in the construction industry of Oman. Out of the 67 responses, 20.9% have experience between 6–10 years, 1.5% have experience between 11–15 years, 19.4% have experience of 16 years and above, and 58.2% have between 1–5 years. However, most respondents (58.2%) have had construction industry work experience in the range of 1–5 years, meaning the majority are fresh graduates and more acceptable to the new technologies. About 44% of the respondents were structural engineers, 30% were architects, and 24% were quantity surveyors. The result is shown in Table 1.

Table 1. Percentages of respondents according to their professional experience period.

Number of years	Frequency	Percent	Cum Percent	Frequency
1–5 Years	39	58.21%	58.21%	58.21%
6–10 Years	14	20.90%	20.90%	79.10%
11–15 Years	1	1.49%	1.49%	80.60%
16- above	13	19.40%	19.40%	100.00%
Total	67	100.00%	100.00%	

When the professionals were asked to evaluate the efficiency and effectiveness of IT to manage information/documents via computers with the appropriate software; 34.8% confirmed that information management is very effective in construction projects. In comparison, 7.6% think it is not practical, as shown in (Fig. 1). This ensures that information management is widely used in electronic versions and is very efficient.

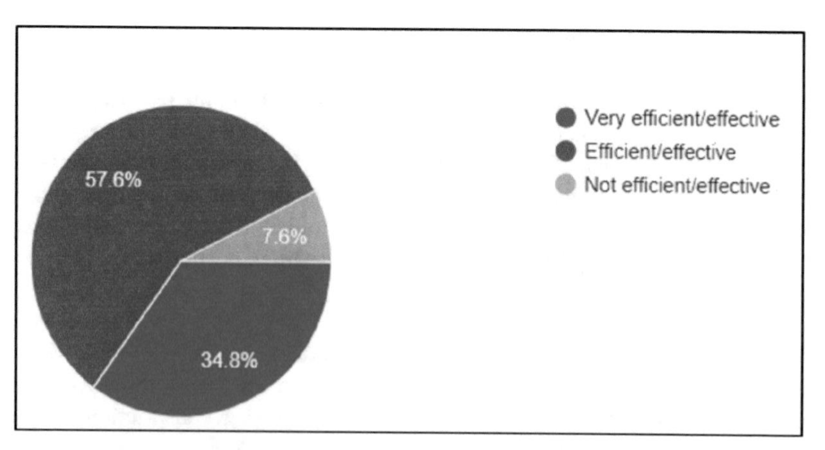

Fig. 1. The extent of the respondent's satisfaction with the use of electronic copies.

The respondents reported that Information and Communication Technology (ICT) could aid their professional duties differently, as shown in Fig. 2. For example, about

42% of them think that IT helps save time, 21% said that IT is highly efficient, 21% said that IT is accurate, and 15% said that IT helps store and retrieve information faster.

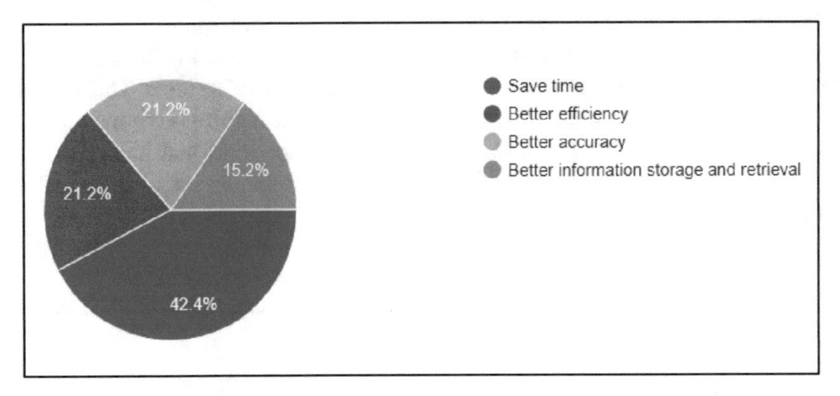

Fig. 2. Ways that information technology helps in the development of the profession.

Respondents were further asked to rate the level of ICT standards in the Oman construction industry as compared to the standards of more developed countries (Fig. 3):

- The survey result showed that 33% agreed that IT is weak in the Sultanate of Oman compared to more developed countries. Furthermore, 52% believed that the presence of IT in the Sultanate of Oman is average. In comparison, 13% thought that IT is better compared to more developed countries.

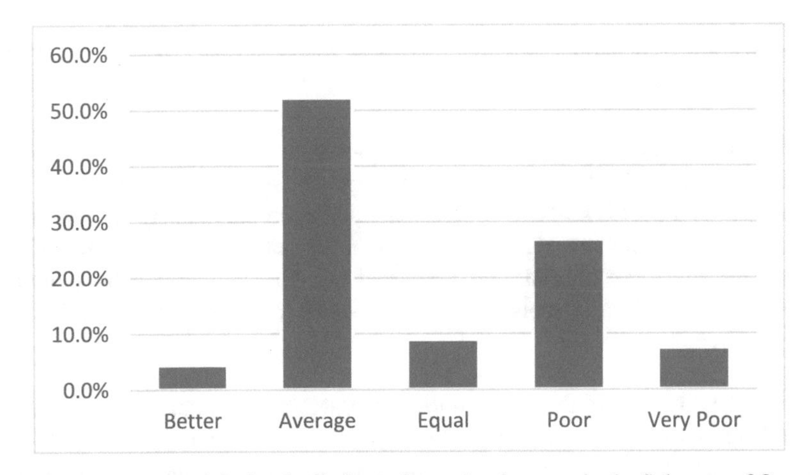

Fig. 3. Assessment of the level of information technology use in the Sultanate of Oman.

Moreover, several factors hinder the use of IT tools. Therefore, the respondents were asked their opinion about the obstacles that impede the use of IT tools. As shown in Fig. 4:

- That 11.9%, think that what prevents them from using technology and information tools is that they are a small percentage of their use in some construction institutions. In contrast, 37.3% say that technology and information tools are not available,
- About 28.4% of respondents said that technology and information tools are very complex, meaning that it is difficult to use them in institutions, and 14.9% said they prefer manual implementation more than technology and information tools. This means that some projects may not need IT and are completed satisfactorily with manual performance only. However, 7.5% said that there are no obstacles that hinder the use of technology and information tools that confirm that technology and information are still used in some construction institutions.

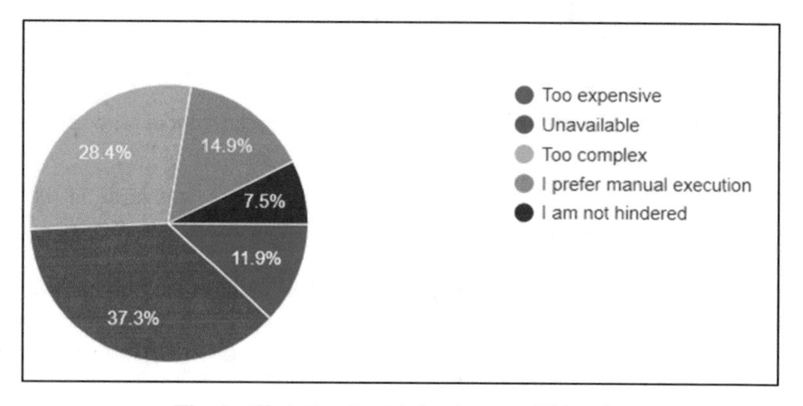

Fig. 4. Obstacles that hinder the use of IT tools.

Furthermore, they were asked how they think they can boost (if need be) the level of IT standards in the Oman construction industry as in Fig. 5.

- We found that about 19.7% thought that more awareness should be created about the importance of IT. While 21.2% of them said that easy-to-use technology should be created, and 36.4% said that professional bodies in the industry should insist on applying new technology. Finally, 16.7% of them said that a cheaper but effective technology should be created, and 1.6% said that the tangible benefits that can be obtained from modern IT tools should be conveyed to the construction industry and its workforce.

Also, the survey found that the majority of construction companies in Oman don't execute international projects. Where 12.3% of the engineers or construction group strongly agree that the company in which they work in execute international projects, and 33.8% only concur that the company execute international projects, while 38.5% strongly disagree that the company execute international projects, and 15.4% do not agree with that and the factors they think hinder the organisations from participating in international projects.

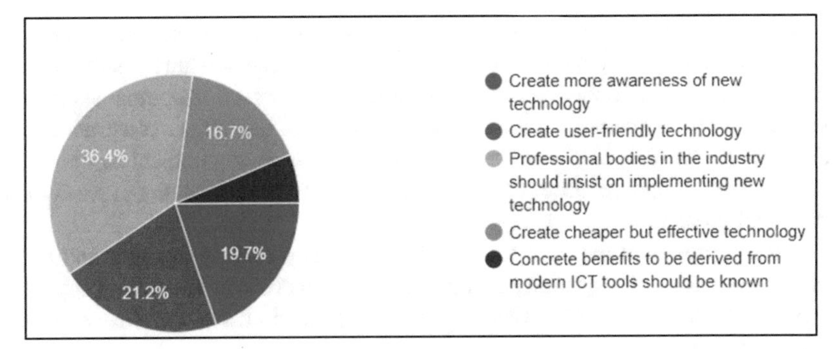

Fig. 5. Practices to enhance IT.

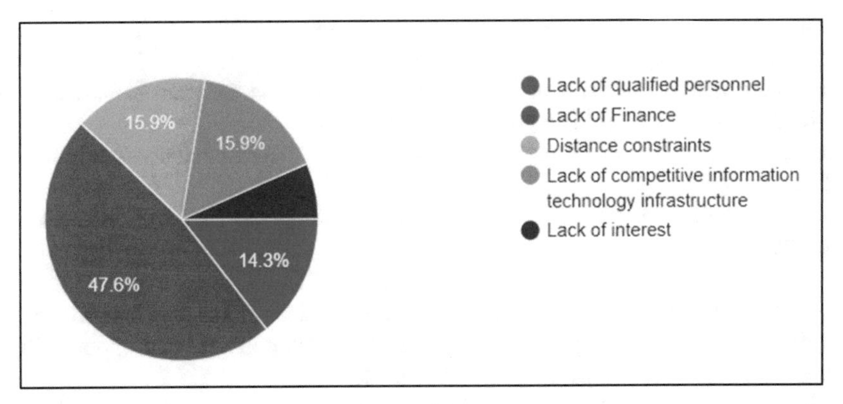

Fig. 6. Factors that hinder the use of information technology in international projects.

According to Fig. 6 shown, 14.3% think that the lack of qualified personnel is a factor that prevents the company from participating in international projects. At the same time, 47.6% of them believed that the lack of finance prevented the company from participating, 15.9% of them agreed that distance constraints were what prevented the company using IT, 15.9% of them assumed that the lack of an IT infrastructure was the factor that prevented the company from participating in international projects.

Finally, 6.3% of them said that the lack of interest was the reason that prevented the company from participating.

4 Conclusion, Recommendations and Future Work

The study aims to assess the extent to which IT is used in the pre-contract stage of construction projects in the Sultanate of Oman. In this paper we demonstrate how successfully IT may improve the construction management process for more effective improvements toward long-term sustainability.

To collect information and data for this study, a questionnaire and interviews with professionals in the ministry of housing and samples of construction companies in Oman were utilised. The study explained in a simple way that IT is of great importance in the construction industry in general and especially in Oman. The study results indicated that IT is widespread in the ministries of housing and construction companies in Oman. However, there is a weakness in the use of IT in some of the Omani construction companies. Due to the lack of information technologies in some companies, they do not participate in electronic bidding (which is more effective), and also do not participate in international projects.

The study presented the obstacles that hinder construction companies and housing ministries from using IT, especially during the pre-contract stage. The two most prevalent obstacles in Omani construction companies were identified, such as the weak capabilities of employees in the use of IT.

To overcome the lack of IT, our study recommends,

- More research to reach a more accessible technology for building designers in general, and quantity surveyors in particular,
- The authorities in Ministry of Housing and construction companies must establish strict laws on the use of IT in construction projects.
- The Ministry of Housing must intensify the work of educational courses in the field of IT in projects,
- All construction companies and housing ministries should have their own effective website.

In conclusion, effective IT implementation for sustainable construction management can only work if multidisciplinary stakeholders see the potential and opportunity for value creation at any point of the project life cycle.

Finally, in this research, we targeted the construction group only, which included civil engineers, architects, and quantity surveyors but didn't include clients; therefore, our future research will include different categories of participant. Moreover, we plan to quantify the effectiveness of using IT technology by measuring the impact of using some IT technologies, such as BIM, in construction projects.

References

1. Plebankiewicz, E., Malara, J.: Analysis of defects in residential buildings reported during the warranty period. Appl. Sci. **10**, 6123 (2020). https://doi.org/10.3390/app10176123
2. Craig, N., Sommerville, J.: Information management systems on construction projects: case reviews (2006). https://www.emerald.com/insight/content/doi/10.1108/09565690610713192/full/html
3. Blessing, M.: E-Agriculture and Rural Development: Global Innovations and Future Prospects: Global Innovations and Future Prospects. IGI Global (2012)
4. Onyegiri, I., Nwachukwu, C.: Information and communication technology in the construction industry. Am. J. Sci. Ind. Res. **2**, 461–468 (2011)
5. Andipakula, T.: A case study of barriers inhibiting the growth of Information and Communication Technology (ICT) in a construction firm. Thesis. Department of Construction Management. Colorado State University Fort Collins, Colorado (2017)
6. Kahura, M.: The Role of Project Management Information Systems towards the Success of a Project (2013). file:///C:/Users/TechRay/Downloads/Documents/10.1.1.677.3317.pdf

7. Castle, G.: New technology-opportunity or threat? In: Cartlide, D. (ed.) New Aspects of Quantity Surveying Practice. ITCon, vol. 2, pp. 1–13 (2002)
8. Stephen, E.P., David, A.M.: Construction Scheduling and Progress Control Using Geographical Information Systems (2006). https://ascelibrary.org/doi/abs/10.1061/(ASCE)0887-3801(2006)20:5(351)
9. Halfawy, M., Froese, T.: Building integrated architecture/engineering/construction systems using smart objects. J. Comput. Civ. Eng. **1a**(2), 172–181 (2005)
10. Klinc, R., Turk, Ž, Dolenc, M.: A service-oriented framework for interpersonal communication in architecture, engineering and construction. Tehnicki vjesnik - Technical Gazette **23**, 1855–1862 (2016)
11. Murray, M., Nkado, R., Lai, A.: The integrated use of information technology in the construction industry. In: Proceedings of the CIB 78 Conference: IT in Construction in Africa, pp. 39-1–39-13. Pretoria, South Africa (2001)
12. Alkalbani, S., Rezgui, Y., Vorakulpipat, C.: ICT adoption and diffusion in the construction industry of a developing economy: the case of the sultanate of Oman (2012). https://www.tandfonline.com/doi/abs/10.1080/17452007.2012.718861
13. Boyce, C., Neale, P.: Conducting In-Depth Interviews: A Guide for Designing and Conducting In-Depth Interviews for Evaluation Input, pp. 2–12. Pathfinder International, Watertown (2006)
14. Stenbacka, C.: Qualitative research requires quality concepts of its own. Manag. Decis. **39**(7), 551–556 (2001). https://doi.org/10.1108/EUM0000000005801
15. Whitley, R.: The Intellectual and Social Organization of the Sciences. Oxford University Press, Oxford (2000)

ROI in Training Projects: From Satisfaction to Business Impact

Leandro F. Pereira[1(✉)], Álvaro L. Dias[1], Rui Vinhas da Silva[1], and Natália L. Teixeira[2,3]

[1] BRU-Business Research Unit, ISCTE - Instituto Universitário de Lisboa, Lisbon, Portugal
{leandro.pereira,rui.vinhas.silva}@iscte-iul.pt

[2] CEFAGE – Centros de Estudos e Formação, Avançada em Gestão e Economia, Évora, Portugal
natalia.teixeira@isg.pt

[3] ISG - Business and Economics School, Lisbon, Portugal

Abstract. While knowledge and methodologies on how to estimate and measure benefits for business initiatives have advanced, it is uncommon to apply this knowledge and methodologies to HR initiatives, particularly training projects. As HR executives need to justify the effectiveness of their training investments and whether they generate any return for employees and the organization's business goals, knowing how to measure ROI in training projects is becoming a critical skill. This study presents the main findings on the current practices of ROI measurement in training projects used by organizations. The study concluded that while levels 1 and 2 (reaction and learning evaluation) are frequently used, the remaining levels (impact, application, and ROI analysis) are often neglected. The article also provides recommendations and directions for future research on the topic.

Keywords: Training Projects · Benefits Realization Management · Business Case · Business Impact · Human Resources · ROI

1 Introduction

The current economic crisis, the rapid improvement of technology, the influence of social media, and market globalization are driving firms to spend more money on professional development. In order to get well-prepared and skilled teams toward the success of the business, the new business problems fueled by the escalating market competitiveness necessitate the creation of new technical and dynamic competences across businesses. Yet, corporate managers, particularly HR executives, are under pressure to maximize their resources and produce positive outcomes (Pereira & Teixeira, 2015).

As a result, the paradigm of the human resources department has shifted. The mission of HR now extends far beyond payroll, firing policies, and holiday planning. Near the Administration Board, HR is playing a crucial function. Today's executives want to know the precise economic return on investment, where every dollar invested should increase the company's bottom line, and are no longer happy with qualitative results

V. Gupta et al. (Eds.): SSEBIM 2022, LNISO 62, pp. 142–152, 2023.
https://doi.org/10.1007/978-3-031-32436-9_12

(Jordo et al., 2020). (Philips, 2007). Every project, project, or initiative implemented in this environment of intense competition and limited resources must create value, or, in other words, more economic return. This can be done in one of three ways: either by increasing business, reducing or avoiding costs, or by improving efficiency (Pereira & Teixeira, 2015).

Therefore, HR executives increasingly need the ability to analyze the return on investment (ROI) in training in order to assess the estimated value of each training project and make the best decisions (Costa et al., 2020). HR will be able to justify the need for qualified candidates, secure the necessary budget approval, and accomplish the ultimate aim of responding to market competition and retaining talent by being aware of the expected return and making the appropriate training investments.

HR should therefore be familiar with and knowledgeable about the following inquiries:

- Is my company aware of the return on investment for the training that was given?
- Are the projects related to training in line with the strategic axes?
- Does the training help the company reach its objectives?
- Do the training performance reports include the proper formal documentation?

Despite increased investments in training human capital, the majority of firms fall short of their goals (Philips, 2007; Philips & Phillips, 2007b).

The following are a few reasons for this return failure:

- Absence of up-front analysis (requires comprehensive assessment) and poor definition of anticipated impacts are two examples of poor business objective definition and training alignment.
- Lack of Manager engagement;
- Inappropriate training project;
- Lack of training follow-up and application evaluation
- Low quality of the training provided
- Undefined methodology

Organizations frequently fail to do in-depth analyses of the issues their employees are currently facing, which would help them determine the skills, knowledge, or behaviors each employee needs to be taught in. The area Directors and HR will be able to decide whether the problem is a training issue or if it has to be addressed in another way by creating a previous diagnosis. Frequently, training projects are given with the expectation that they will lessen certain negative effects when they fail to produce those results because a non-training solution would be the best option (such as compensation strategies, team building, managerial coaching, process and workflow improvements, equipment repair or replacement, workload adjustment, and others) (Philips & Phillips, 2007b).

Having said that, it is becoming increasingly important to invest resources in the correct training projects and to prevent squandering money on training sessions that are just necessary for legal compliance. No one is that ignorant, in the words of Blaise Pascal, who has nothing to teach. Nobody is truly wise if they haven't learned anything.

2 Literature Review

Corporate learning is not new; one instance is on-the-job training, in which a new employee was trained by an experienced employee (Devarakonda, 2019) and its effects may be seen in all aspects of a business, including its financial outcomes and the motivation and dedication of its employees (Jasson and Govender, 2017). As a result, it is becoming more important to assess the success of training (Srimannarayana, 2017). Companies utilize training to alter their employees' behavior, knowledge, and skills in order to improve performance and help them accomplish their goals (Jasson and Govender, 2017). It is unclear how firms can determine the return on investment (ROI) of these processes, even while they recognize the value of increasing employee abilities to remain competitive and keep employees motivated (Jasson and Govender, 2017). Even while the value of training is widely acknowledged, many executives still struggle to justify the money they spend on staff training, particularly in industries like stock investing and sales where training frequently yields variable outcomes. Also, the high rate of employee turnover and the subpar quality of the training that is offered are two variables that reduce the investments made in corporate training (Devarakonda, 2019). Gaining a stronger return on investment from training will depend on identifying and assessing these risks, then managing them with consideration for people development and business performance (Jasson and Govender, 2017).

Today, the managerial team as a whole must support and participate in the evaluation of training because it is no longer solely the responsibility of the human resources department (Srimannarayana, 2017). Nevertheless, most managers today lack the skills and resources necessary to assess the value and Return on Investment (ROI) of training. This influences how leaders make decisions, making them less willing to spend money on training since they don't think it will yield a good return. The effect that training may have on the organization's financial sheet is one way that return can be assessed.

As a result, businesses must begin tracking the time, money, and resources used in training projects, viewing them as investments rather than expenses and measuring their financial (ROI) impact (Devarakonda, 2019). Teams in charge of training will be able to convincingly demonstrate that training is a factor that will result in measurable benefits and greatly improve a company's financial performance if they can create a thorough training evaluation procedure (Srimannarayana, 2017).

In another light, it's critical to stress that when a business doesn't achieve the training impact anticipated, it's necessary to comprehend the cause and use a problem-solving method (Pereira & Santos, 2020). This suggests that a knowledge base be created to enable ongoing learning on the successes and failures of earlier training efforts (Pereira et al., 2021).

Almost 5,000 firms employ the ROI Methodology, which is regarded as the most widely used and applied evaluation system worldwide (Philips, 2007). The ROI Methodology enhances project design for maximum impact in addition to giving the opportunity to assess project performance. The desired objectives for each level of the framework are listed below, along with some collecting technique ideas (which must be suitable for the analysis's goal) (Philips & Phillips, 2007a; Kaminski & Lopes, 2009) (Fig. 1):

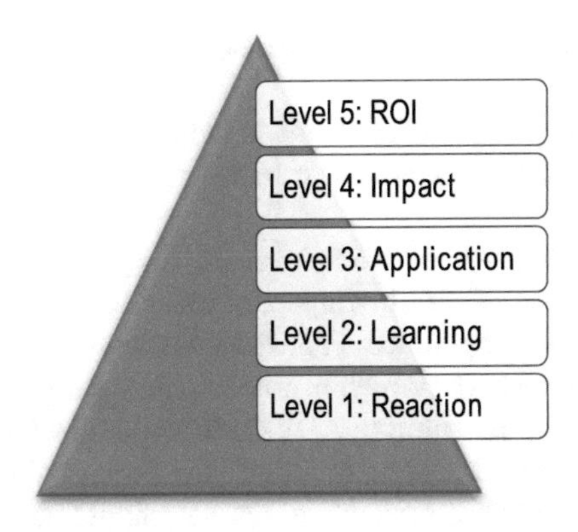

Fig. 1. ROI Framework. Source: Adapted from Philips (2007)

Level 1: **Reaction**: analyze the participants' responses to the project. What were they fond of? What do they think about the schedule, environment, and trainer? Did they think the training was worthwhile?

Techniques: Survey; Interviews.

Level 2: **Learning**: measures changes in abilities, knowledge, or attitude. (Change in behavior, abilities, and knowledge).

Techniques: Pre and Post-training assessments eg. Test performance, role play, pre-post test.

Level 3: **Application**: evaluates specific application and implementation as well as behavioral changes that occur at work. (In work, changing how things are done).

Techniques: Observation; Interview.

Note: Adaptation requires time. To compare, this analysis should be conducted both before and after training.

Level 4: Impact: evaluates the effect on business (project benefits). Overall improvement to the company (e.g., task costs are lower, fewer workplace accidents occur, sales are higher, and efficiency is higher).

Techniques: Control Group; Trend Analysis.

In order to connect training with business-unit measurements, sponsors of training and development (the major clients who fund and support the project) often demand that Level 3 and Level 4 objectives clearly demonstrate how the training project's outcomes will truly benefit the company. These objectives and associated indicators need to be gathered following the training session in order to compare the estimated training benefits with the actual benefits realized.

Level 5: Return on Investment: compares the project's financial gains to its costs, expressed as a percentage. Examine how the bottom line has altered, and determine whether the advantages outweigh the disadvantages. We must gather information through assessment and evaluation on the knowledge and skills that were acquired as well as the behaviors that changed in order to calculate ROI. If the behavior of the employees does not change, there will be no impact and no ROI (Philips, 2007). Pereira Diamond advises identifying the benefits within four dimensions to determine the initiative's impact: company growth, cost reduction, efficiency improvement, or legal compliance (Pereira & Teixeira, 2015).

The primary advantages are listed below to help you comprehend the ROI contribution to Management teams (Kirkpatrick & Kirkpatrick, 2005; Philips, 2007; Philips & Phillips, 2007b):

Measure Contribution – It is feasible to obtain the precise contribution from a number of projects, or in other words, the impact of learning in numerical terms, by adhering to an exact and reliable approach. A ROI analysis will show whether the project's advantages outweighed its costs, which is crucial data for improved HR management;

Establish Priorities – By figuring ROI Understanding which initiatives contribute the most in various areas enables one to choose the projects that will have the greatest influence on the firm. The least effective initiatives should be abandoned, whereas the most inefficient ones could be reworked or redeployed;

Focus on Results – This technique is an outcomes-based process, therefore it places emphasis on the project results and calls for focus on quantifiable goals, or what the project aims to achieve, from all parties involved;

Earn respect of Senior Executives and Sponsor – Senior executives want to know how actions will affect their return on investment, therefore they will value initiatives to link training to business implications and quantify benefits;

Alter Management Perceptions of Learning and Development – enables upper management to comprehend that education is an investment rather than a cost.

3 Methodology

This study aims to gain a deeper understanding of the ROI in Learning techniques used by HR and identify the primary areas that require improvement in light of the growing need to improve training investment strategies in enterprises. HR professionals (Directors, Managers, and Coordinators) were contacted and invited to take part in a survey from a variety of industries, including banking, construction, energy, retail, health, insurance, telecommunications, and others. The researchers utilized a convenience sample since they only wanted people who understood the significance of the topic.

Six Likert-scaled multiple-choice questions and one open question made up the survey, which was created using best practices from the ROI Methodology. 53 people responded to 100 surveys that were sent out over a two-week period in May 2019; which is a response rate of 53%. In terms of the sample profile, 26% of participants were HR Directors, 40% were HR Managers, and 33% were HR coordinators. 78% of people were male and 22% were female. Regarding the organization dimension in terms of the number of employees, 11% have fewer than 10, 9% have between 10 and 49, 49% have between 50 and 250, and 30% have more than 250.

4 Result Analysis

Several conclusions can be drawn from the responses. The survey's questions and the associated data analysis are shown below (Fig. 2).

This query examines the bottom level of the ROI Pyramid.

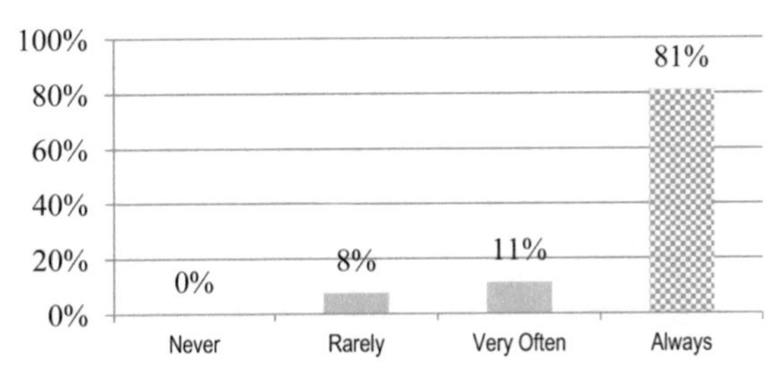

Fig. 2. Do you evaluate the participants **satisfaction** after each training program is completed? (eg. Questionnaires). Source: Self-constructed

The participant's reaction evaluation is the most frequently used of the five levels: 81% of participants say they always use it, 11% say they use it frequently, and only 8% say they seldom use it. These figures are quite encouraging since they demonstrate that reaction analysis is a problem that most firms share. Yet, do businesses take into account participant input to enhance the caliber and efficacy of trainings? Is the reason for any unfavorable classification of the training session questioned? These are essential factors to take into account for an effective satisfaction assessment. Even while the response to the project may have been outstanding, this does not necessarily mean that there has been no effect (no changes in results) or learning transfer (application on the job) (Kaminski & Lopes, 2009) (Fig. 3).

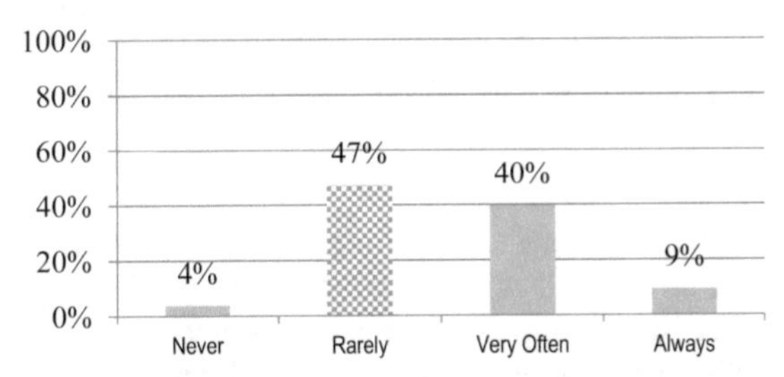

Fig. 3. Do you evaluate the **knowledge** and competencies from participants, both prior and after the program training session? Source: Self-constructed

This inquiry examines the learning gains made by training, which makes up the second level of the ROI Pyramid. At this point, it is feasible to see a significant decrease from the prior level. Just 9% of respondents recommend doing it always, 40% recommend doing it frequently, 47% recommend doing it infrequently, and 4% have never finished this work. According to these statistics, approximately half of the company fails to complete the learning evaluation process immediately following the training project, making it impossible for them to determine the extent to which it benefited the intended audience (knowledge or skills improvement). Participants were asked if they created an application plan at the end of the project and if they created a systematized follow-up to validate it in order to assess the third level (Fig. 4).

Only 6% of respondents to the question of how often they prepare an application plan said Always, 25% said Very Often, and 56% said Sometimes. 13% of people say

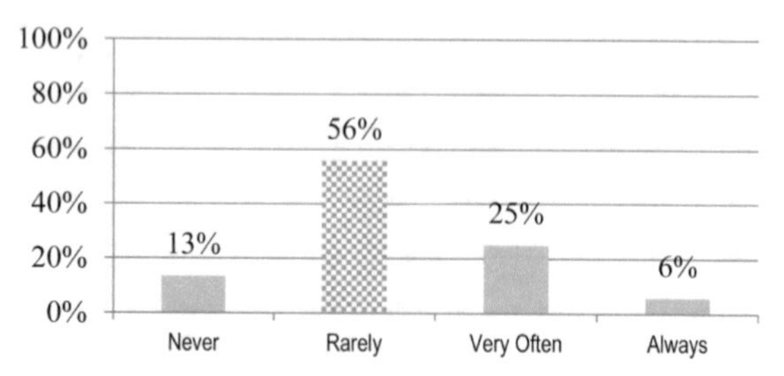

Fig. 4. Do you make an **application** plan for the technical competencies and tools teached in the training session? Source: Self-constructed

they never do it. These figures imply that the majority of businesses do not create any plans at the conclusion of the training project that detail how they will use the newly gained information and abilities in their regular operations (daily tasks) (Fig. 5).

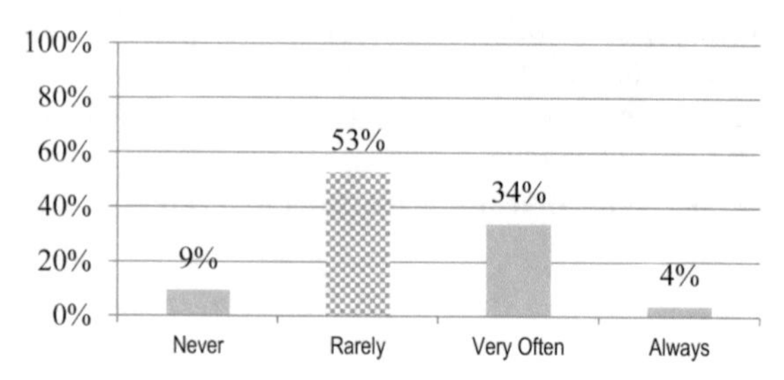

Fig. 5. Do you make a **sistematized follow-up** after training to validate the application? Source: Self-constructed

A follow-up is not usually done, according to only 4% of respondents, 34% say it is done frequently, 53% say it is done infrequently, and 9% say it is never done. These outcomes are the result. A systematized approach to verifying the application of new techniques, skills, and information obtained will be hampered if there is no prior application strategy for the tasks, projects, or processes to which they will be applied. How can we determine whether training was effective if there is no follow-up in the employee's actual work environment? If an employee's behavior does not change, there will be no impact, no advantages, and no good ROI. This information must therefore be understood in order to better management decisions. If this phase is skipped, it will be hard to create an impact study and calculate the exact worth of the advantages brought about. In other words, it will be impossible to determine whether the firm benefited from the investment in training (Fig. 6).

The data indicates that 11% confirms a quantitative impact measurement, and 4% consistently conducts an application analysis. Yet, can an impact be measured without monitoring the application? Are people continually aware of and knowledgeable about what ROI entails since there cannot be a ROI without an application?

While deciding whether to pursue implementing the training project or not, HR and the Area Director were required to gather information about the present issues/challenges that their employees (team) were facing and how those issues were affecting them on a

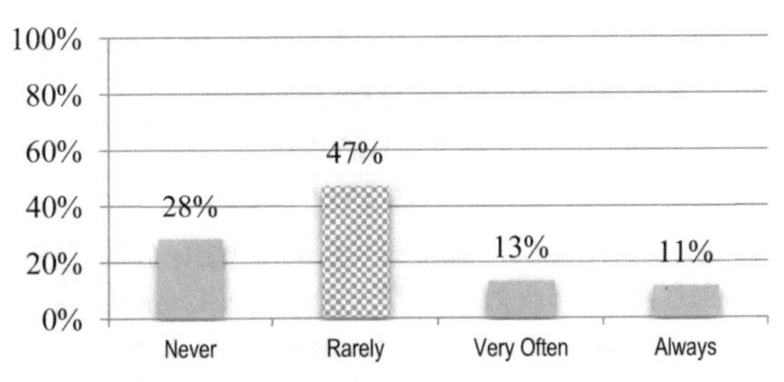

Fig. 6. Is the training **quantitative impact** measured during the employees activity? (eg. Metrics, KPIs). Source: Self-constructed

daily basis. Before making a selection about the training project solution, it is essential to determine which tasks or processes need to be enhanced and comprehend how they relate to the business results. After having the current situation (before to training) clearly recognized (i.e. documented using metrics), the same metrics should be validated and used to determine whether and how much progress there was. We should be able to respond to the following queries in this step:

- Was the training successful?
- Did it correspond to the previously noted "pains"?
- How much money those benefits are worth in terms of value?

The organization should develop specific metrics and kpis that "image" the current situation (before to the training) and ensure its collection once more, after the training project, to help with acquiring this information (Fig. 7).

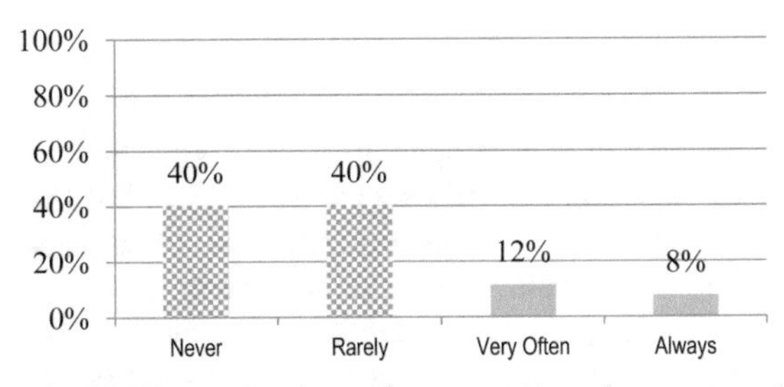

Fig. 7. Is there any **evaluation** to measure the economic value generated by each training program? Source: Self-constructed

The final phase seeks to determine whether the investment contributed value, or, in other words, if the benefits leveraged (positive business impacts) were higher than the expenditure, once the training benefits have been recognized and financially measured. The costs are often made up of both fixed and variable charges. No matter how many participants there are in the training project, the fixed costs are the ones that stay the same (Kaminski & Lopes, 2009). On the other side, variable costs are those that vary depending on who attends, such instruction manuals, breaks for food and coffee, travel expenses, training fees, and many more. The return on investment (ROI) indicator, which should be at least a positive amount (ROI > 0), is advised for determining the percentage of value contributed. The ROI calculation is shown below:

$$\text{ROI}(\%) = \frac{\text{Net HR Project Benefits}}{\text{HR Project Costs}}$$

This inquiry enables drawing the alarming conclusion that 40% of the organizations just occasionally compute ROI and another 40% never did. Organizations won't be able to compute ROI if they neglect to gather training impacts (benefits).

It is crucial to remember that some trainings, such as those focused on behavior (e.g., leadership, teamwork, attitude, and a more positive workplace environment), can only be evaluated over time (Kaminski & Lopes, 2009). People frequently change their habits right away after receiving training, but over time they have a tendency to revert to their old habits, so these benefits assessments should be conducted after the training project has been completed for two to three months, and they should be repeated at six months. After training, change frequently occurs, but employees quickly fall back into old routines or practices, or they may even feel pressure from peers or coworkers to resist the change. The benefits assessments should be conducted 2–3 months after the training project and again at 6 months out for this reason.

5 Research Limitations

Considering the research's limitations, it is advised that future studies explore using a bigger sample size because it will statistically produce results that are more diverse and representative of the population. The sample that was examined was small and based on convenience sampling. It would have been fascinating to conduct a factor analysis for a more in-depth study, as this could have produced more illuminating data (for instance, compare the results between the business sectors; business sizes; HR roles, and others)..

6 Conclusion

In conclusion, despite the highly developed processes for business cases and return on investment, HR departments continue to struggle with completing the entire process for training initiatives. The results of this investigation show, among other things:

The participant's reaction evaluation is the most common practice among the five levels; the learning evaluation process is missed by half of the organizations immediately after the training project; the majority of companies do not create any plans at the conclusion of the training project; the systematized follow-up after training to validate the application is a limited option for the majority of respondents; and the ROI is infrequently calculated by 40% of the organizations.

Organizations that make poor judgments not only squander resources (money, time, and human capital), but also jeopardize their ability to fulfill corporate goals and run the danger of losing important, talented personnel (Fig. 8).

Figure 8 shows a barometer within the ROI pyramid where we may determine the maturity level of the implemented process based on how frequently each procedure is used. In conclusion, human resources executives who play a strategic role alongside top management must have the ability to accurately estimate the advantages of training projects and make sound decisions.

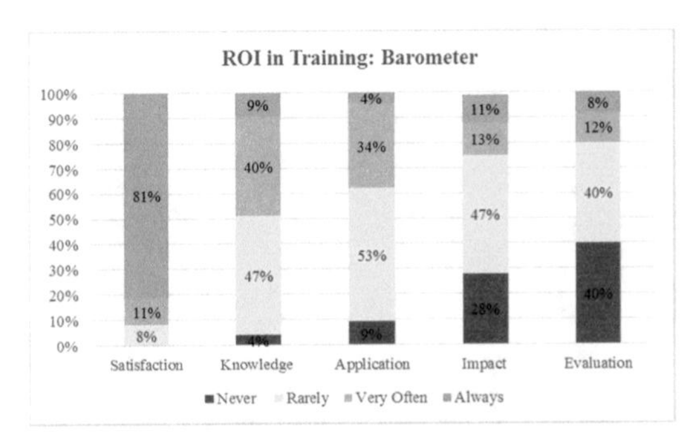

Fig. 8. ROI in Training Barometer. Source: Self-constructed

References

Costa, R., Resende, T., Dias, A., Pereira, L., Santos, J.: Public sector shared services and the lean methodology: implications on military organizations. J. Open Innov.: Technol. Mark. Compl. **6**(3), 78 (2020)

Devarakonda, S.: Calculating the economic viability of corporate trainings (Traditional & eLearning) using benefit-cost ratio (BCR) and return on investment (ROI). Int. J. Adv. Corpor. Learn. **12**(1), 41–57 (2019). https://doi.org/10.3991/ijac.v12i1.9735

Jasson, C.C., Govender, C.M.: Measuring return on investment and risk in training – a business training evaluation model for managers and leaders. Acta Commercii **17**(1), 1–9 (2017). https://doi.org/10.4102/ac.v17i1.401

Jordão, A.R., Costa, R., Dias, Á.L., Pereira, L., Santos, J.P.: Bounded rationality in decision making: an analysis of the decision-making biases. Bus.: Theory Pract. **21**(2), 654–665 (2020)

Kaminski, K., Lopes. T.: Society for human resource management (2009)

Kirkpatrick, D., Kirkpatrick, J.: Transferring Learning to Behavior: Using the Four Levels to Improve Practice. Barrett-Koehler, San Francisco, CA (2005)

Pereira, L., Teixeira, C., Salgado, A.: Pereira diamond: projects' economic and social impacts. In: 23rd International Conference on Engineering, Technology and Innovation, ICE/ITMC 2017, pp. 6–14. IEEE, Funchal (2017)

Pereira, L., Santos, J.: Pereira problem solving. Int. J. Learn. Chang. **12**(3), 274–283 (2020)

Pereira, L.; Da Costa, R.; Dias, A.; Santos, J.: Knowledge management in projects. Int. J. Knowl. Manag. **17**(1) (2021). 1

Philips, J.: Measuring ROI: The Process, Current Issues and Trend (2007). www.roiinstitute.net

Philips, J., Phillips, P.: Show me the money (2007a). www.roiinstitute.net

Philips, J., Phillips, P.: Eleven Reasons why training and Development Fails (2007b). www.roiinstitute.net

Srimannarayana, M.: From reactions to return on investment: a study on training evaluation practices. Indian J. Ind. Relat. **53**(1), 1–20 (2017)

Author Index

V. Gupta et al. (Eds.): SSEBIM 2022, LNISO 62, p. 153, 2023.
https://doi.org/10.1007/978-3-031-32436-9

Printed in the United States
by Baker & Taylor Publisher Services